Artificial Intelligence and Data Mining in Healthcare

Malek Masmoudi • Bassem Jarboui • Patrick Siarry
Editors

Artificial Intelligence and Data Mining in Healthcare

Editors
Malek Masmoudi
LASPI, G037
IUT de Roanne
Roanne, France

Bassem Jarboui
Department of Business
Higher Colleges of Technology
Abu Dhabi, United Arab Emirates

Patrick Siarry
Laboratoire LiSSi (EA 3956)
Université Paris-Est Créteil Val-de-Marne
Créteil, France

ISBN 978-3-030-45242-1 ISBN 978-3-030-45240-7 (eBook)
https://doi.org/10.1007/978-3-030-45240-7

This Springer imprint is published by the registered company Springer Nature Switzerland AG.
The registered company address is: Gewerbestrasse 11, 6330 Cham, Switzerland

Preface

Our healthcare systems are facing unprecedented challenges and particularly serious economic pressure. Promising alternatives to the traditional healthcare system are being developed to improve the quality of service and to reduce cost. Researchers in the operations research and artificial intelligence communities are involved to provide efficient healthcare decision support systems to help healthcare engineers and managers to make optimal, efficient decisions. How to improve the whole healthcare system performance has become the main issue for researchers from medical, technical, organizational, and decisional points of view. The need for intelligent support systems for decision-making is growing in different healthcare fields. The domain is complex and very rich in terms of scientific niches which attract researchers in both the operations research (OR) and artificial intelligence (AI) disciplines.

This book, "Artificial Intelligence and Data Mining in Healthcare," presents recent studies and work in healthcare management and engineering using artificial intelligence and data mining techniques. It focuses on mainly exposing readers to cutting-edge research and applications that are ongoing across the domain of healthcare management and engineering where artificial and data mining techniques can be and have been successfully employed.

Need for a Book on the Proposed Topics

To the best of our knowledge, there is no book aiming precisely at regrouping artificial intelligence and data mining techniques for healthcare decision-making problems, and the number of book chapters dedicated to this subject is tiny. However, this topic is highly topical and interests many researchers, which explains the high number of journal papers and international conferences communications dedicated to this subject.

This provides readers with AI and data mining tools for solving healthcare decision-making problems. It explains a wealth of both basic and advanced concepts of AI and data mining applied to organizational tasks such as patient workflow, capacity/resource management, logistics, medical image compression, life expectancy, etc. The chapters include relevant case studies.

Organization of the Book

This book is organized into nine chapters. A brief description of each chapter is given below.

The chapter "Artificial Intelligence for Healthcare Logistics: An Overview and Research Agenda" by M. Reuter-Oppermann and N. Kühl examines the existing literature on artificial intelligence and machine learning approaches for the logistical problems that arise when we design, provide, and improve healthcare services. For the analysis, we distinguish between the planning levels (strategic, tactical, and operational), the care level (primary, secondary, and tertiary care), and the resource types (doctors, nurses, technicians, patients, etc.). Based on the results we provide a research agenda with open topics and future challenges.

The chapter "Synergy Between Predictive Mining and Prescriptive Planning of Complex Patient Pathways Considering Process Discrepancies for Effective Hospital-Wide Decision Support" by T. Mellouli and T. Stoeck considers decision making and decision support tasks for planning complex patient-centered clinical pathways in a complex hospital environment, demarcated by many wards, shared resources, and many other interdependencies. Introducing a two-dimensional scheme with these complexity dimensions, many AI- and OR-oriented tasks in hospitals are classified and several facets of AI/OR synergy for their effective solution are detected. The first type of AI/OR synergy forwards process mining results of complex pathways (AI) to prescriptive optimization models (OR). Case studies of a university hospital show business benefits and better results quality. Based on a profound discussion, a second hidden type of AI/OR synergy is detected, where hard-to-model interdependencies can be taken into consideration. The operationalization of this AI/OR synergy is based on a procedure for discrepancy mining (AI) which is embedded with a prescriptive model (OR) into a plan-and-refine framework.

The chapter "Real-Time Capacity Management and Patient Flow Optimization in Hospitals Using AI Methods" by J.R. Munavalli, H.J. Boersma, S.V. Rao, and G.G. van Merode demonstrates how optimization models based on modern artificial intelligence (AI) techniques would manage hospital workflow through decision-making systems that are dynamic, robust, and real-time. In particular, multi-agent systems and ant colony optimization have the potential to convert traditional workflow management into intelligent and efficient workflow management systems that improve hospital performance. The application of AI to operations management is demonstrated with examples from hospitals.

The chapter "How Healthcare Expenditure Influences Life Expectancy: Case Study on Russian Regions" by N. Mladenovic, O. Rusetskaya, S. Elleuch, and

B. Jarboui examines the influence of healthcare supports of different kinds on life expectancy. Data are collected on all 85 geographical districts in Russia, covering a 15-year period. A symbolic regression model is applied and solved using variable neighborhood programming, a recent promising automatic programming technique. In other words, the analytic function is searched to present the relationship between life expectancy and a few selected healthcare financial attributes. Some years are used as a training set and some as a testing set. Interesting results are obtained and analyzed. They confirm the fact that symbolic regression and artificial intelligence techniques might be the right approach to estimating life expectancy.

The chapter "Operating Theater Management System: Block-Scheduling" by B. Bou Saleh, G. Bou Saleh, and O. Barakat deals with block-scheduling in operating theater management using a MILP and a distributed artificial intelligence approach. The provided approach takes into consideration variations in doctor and operating room availabilities. A real case study is considered to make simulations that show the superiority of the distributed artificial model in comparison with MILP.

The chapter "An Immune Memory and Negative Selection to Visualize Clinical Pathways from Electronic Health Record Data" by M. Berquedich, O. Kamach, M. Masmoudi, and L. Deshayes provides a data-driven clinical practice development methodology to extract common clinical pathways from patient-centric electronic health record data. An algorithmic methodology is proposed to handle this type of routine data. In this chapter, the authors design a system of control and analysis of patient records based on an analogy between the elements of the new electronic health records (EHR) and biological immune systems. The detection of patient profiles is handled using bi-clusters. The authors rely on biological immunity to develop a set of models for structuring knowledge about EHR and pathway analysis decisions. A specific analysis of the functional data led to the detection of several types of patients who share the same information on their EHR. This methodology demonstrates its ability to simultaneously process data and provide information for understanding and identifying the path of patients as well as predicting the path of future patients.

The chapter "Optimized Medical Image Compression for Telemedicine Applications" by K.M. Hosny, A.M. Khalid, and E.R. Mohamed provides an algorithm for highly efficient compression of 2D medical images that use Legendre moments to extract features and differential evolution (DE) to select which of these moments are the optimum. The proposed algorithm aims to achieve the best-reconstructed image quality. Medical images from different imaging modalities such as magnetic resonance imaging (MRI), computed tomography (CT), and X-ray Images are used in testing the proposed algorithm. The mean square error (MSE), peak signal-to-noise ratio (PSNR), structural similarity index measure (SSIM), and normalized correlation coefficient (NCC) are quantitative measures used to evaluate the performance of the proposed algorithm and well-known existing medical image compression methods. The quality of the reconstructed compressed images using the proposed method is much better than those compressed using conventional 2D compression algorithms.

The chapter "Online Variational Learning Using Finite Generalized Inverted Dirichlet Mixture Model with Feature Selection on Medical Data Sets" by M. Kalra and N. Bouguila provides a statistical framework for online variational learning using the finite generalized inverted Dirichlet mixture model for clustering medical images data by simultaneously using feature selection and image segmentation. The model allows one to adjust the mixture model parameters, the number of components, and the feature weights to tackle the challenge of overfitting. The algorithm in this study has been evaluated on synthetic data as well as in three medical applications for brain tumor detection, skin melanoma detection, and computer-aided detection of malaria.

The chapter "Entropy-Based Variational Inference for Semi-bounded Data Clustering in Medical Applications" by N. Manouchehri, M. Rahmanpour, and N. Bouguila considers inverted Dirichlet mixture models for semi-bounded positive vectors clustering. An entropy-based variational approach is developed to test whether each component is truly distributed as an inverted Dirichlet. To accomplish this goal, the theoretical entropy of each component is compared with the estimated one and the component with the maximum differential is chosen to be split. The provided approach will be tested in real-world medical applications and compared with the conventional techniques.

Audience

This book is valuable for researchers and master's and PhD students in departments of computer science, information technology, industrial engineering, and applied mathematics, and in particular for those engaged with AI and data mining topics in the healthcare domain.

Roanne, France — Malek Masmoudi
Abu Dhabi, UAE — Bassem Jarboui
Créteil, France — Patrick Siarry

Contents

1 Artificial Intelligence for Healthcare Logistics: An Overview and Research Agenda 1
Melanie Reuter-Oppermann and Niklas Kühl
1.1 Introduction 1
1.2 Machine Learning and Artificial Intelligence 2
1.2.1 Machine Learning 2
1.2.2 Artificial Intelligence 3
1.2.3 Working Definition 4
1.3 Framework for Healthcare Logistics Literature 5
1.3.1 Planning Levels 5
1.3.2 Care Levels 6
1.3.3 User Types 6
1.3.4 Framework 7
1.4 Literature Review 8
1.4.1 AI for Optimisation Input 9
1.4.2 AI for Healthcare Logistics Optimisation 11
1.4.3 AI for ED Logistics 13
1.4.4 Synthesis and Research Agenda 15
1.5 Conclusion 15
References 16

2 AI/OR Synergies of Process Mining with Optimal Planning of Patient Pathways for Effective Hospital-Wide Decision Support 23
Taïeb Mellouli and Thomas Stoeck
2.1 Motivation and Research Outline 23
2.1.1 AI/OR Synergies meet Hospital Decision Task Complexities 24
2.1.2 Pathway Centered Decision Support Toward AI/OR Synergy 25
2.1.3 Research on AI/OR Synergy and Chapter Outline 27

2.2 First Type of AI/OR Synergy: Process Mining of Pathways for Accurate Prescriptive Planning of Ward-and-Bed Allocation..... 28
2.2.1 Synergy between Predictive and Prescriptive Analytics: Cases of Simple vs. Complex Structures 28
2.2.2 First Type of AI/OR Synergy and Its Benefits for Effective Hospital Decision Support: Case Study of a University Hospital 30
2.3 Detecting AI/OR Synergies Within Hospital Decision Support: Interdependencies, Dimensions of Complexity, Two-Dimensional Scheme, and Types of AI/OR Synergy 33
2.3.1 Types of Interdependencies: First Group 33
2.3.2 Dimensions of Complexity and Overview About OR and AI Tasks and Synergies 36
2.3.3 A New Two-Dimensional Scheme for Simulation-/Optimization-Based Decision Support in Hospitals Applied to Overall Bed Management in Interdependent Wards......................... 38
2.3.4 AI Tasks and AI/AI Synergy: Stepwise Aggregation from Process Mining to More Accurate Hospital Data Mining.. 41
2.3.5 OR Tasks and OR/OR Synergies............................... 42
2.3.6 First Type of AI/OR Synergy and Detecting a Second Type.. 43
2.4 Second Type of AI/OR Synergy: Mining of Process Discrepancies and Its Interplay with Prescriptive Planning Toward Effective Hospital-Wide Decision Support 46
2.4.1 Types of Interdependencies: Second Group and Model–Reality Gap 46
2.4.2 Mining Process Discrepancies by Type of Interdependency 49
2.4.3 Interplay Between Mining Process Discrepancies with Prescriptive Planning and Operationalization of the Second Type of AI/OR Synergy by a Discrepancy-Driven Approach............................ 50
2.5 Conclusion 52
References 53

3 Real-Time Capacity Management and Patient Flow Optimization in Hospitals Using AI Methods............................. 55
Jyoti R. Munavalli, Henri J. Boersma, Shyam Vasudeva Rao, and G. G. van Merode
3.1 Introduction 56
3.2 Capacity Management in Hospitals 56
3.2.1 Traditional Hospital Capacity Management.................. 56
3.2.2 Queuing and Synchronization in Hospitals.................... 57

3.3 AI Methods for Hospital Capacity Management 59
3.3.1 Multi-Agent Systems 60
3.3.2 Artificial Neural Networks (ANN) 61
3.3.3 Machine Learning (ML) 61
3.4 Example of AI in Capacity Management and Patient Flow Optimization 61
3.4.1 Methods 62
3.5 Conclusion 65
3.6 Future of AI in Patient Flow Optimization and Capacity Management 65
References 65

4 How the Health-Care Expenditure Influences the Life Expectancy: Case Study on Russian Regions 71
Nenad Mladenovic, Olga Rusetskaya, Souhir Elleuch, and Bassem Jarboui
4.1 Introduction 71
4.2 Life Expectancy as a Symbolic Regression Problem 73
4.3 Variable Neighborhood Programming for Solving Symbolic Regression Problem 74
4.4 Case Study on Life Expectancy at Russian Districts 75
4.4.1 One-Attribute Analysis 76
4.4.2 Results and Discussion on Three-Attribute Data 77
4.5 Conclusions 81
References 82

5 Operating Theater Management System: Block-Scheduling 83
Bilal Bou Saleh, Ghazi Bou Saleh, and Oussama Barakat
5.1 General Context 84
5.1.1 Introduction 84
5.1.2 CHU Operating Theater “Block-Schedule” 84
5.1.3 Search Background 85
5.1.4 Problem Definition 85
5.2 MILP Problem Formulation 86
5.2.1 Definition of Decision Variables 86
5.2.2 Objective Function 86
5.2.3 Constraints 87
5.2.4 Results of the Simulation 87
5.3 MAS Planner Approach 89
5.3.1 Preface 89
5.3.2 Multi-Agent Planner 89
5.3.3 Patient’s Programming 91
5.3.4 Frequency Evaluation 91
5.3.5 Surgeon’s Preferences 92
5.3.6 Virtual Cost 92
5.3.7 Block-Scheduling Algorithm 93

5.3.8 Optimization Algorithm 94
5.3.9 Performance Metrics 94
5.4 Experimental Test 95
5.4.1 Test Data 95
5.4.2 Simulation Results 96
5.5 Conclusion 98
References 98

6 An Immune Memory and Negative Selection to Visualize Clinical Pathways from Electronic Health Record Data 99
Mouna Berquedich, Oulaid Kamach, Malek Masmoudi, and Laurent Deshayes
6.1 Introduction 100
6.2 What is an EHR? 101
6.2.1 Benefits with the EHR 101
6.2.2 Better Practice Management with the EHR 102
6.3 Current Study 105
6.4 Overview of the System 107
6.4.1 Representation of the Self-Cell 107
6.4.2 Antigen Representation 108
6.4.3 Representation of B-Cells 109
6.5 Negative Selection Algorithm for System Monitoring 110
6.5.1 Step 1: Learning 110
6.5.2 Step 2: Monitoring 111
6.6 Control of EHRs by Memory Cells 112
6.6.1 The Algorithm Developed by Immune Memory (IMA) 112
6.7 Implementation and Results 113
6.8 Conclusion 115
References 115

7 Optimized Medical Image Compression for Telemedicine Applications 119
Khalid M. Hosny, Asmaa M. Khalid, and Ehab R. Mohamed
7.1 Introduction 119
7.2 Previous Approaches 120
7.3 Preliminaries 122
7.3.1 Legendre Moments 122
7.3.2 Whale Optimization Algorithm (WOA) 123
7.4 The Proposed Compression Method 126
7.5 Numerical Experiments 129
7.5.1 Test image 129
7.5.2 Performance Measures 129
7.5.3 Results and Discussion 131
7.6 Limitations of the Proposed Algorithm 139
7.7 Conclusion 140
References 140

8 Online Variational Learning Using Finite Generalized Inverted Dirichlet Mixture Model with Feature Selection on Medical Data Sets ... 143
Meeta Kalra and Nizar Bouguila
8.1 Introduction ... 144
8.2 Clustering Applications in Healthcare ... 147
8.3 Model Specification ... 147
8.3.1 Finite Generalized Inverted Dirichlet Mixture Model with Feature Selection ... 148
8.3.2 Prior Specifications ... 150
8.4 Online Variational Learning for Finite Generalized Inverted Dirichlet Mixture Mode with Feature Selection ... 151
8.4.1 Variational Inference ... 152
8.4.2 Online Variational Inference ... 155
8.5 Experimental Results ... 162
8.5.1 Image Segmentation ... 162
8.5.2 Synthetic Data ... 163
8.5.3 Medical Image Data Sets ... 165
8.6 Conclusion ... 171
References ... 174

9 Entropy-Based Variational Inference for Semi-Bounded Data Clustering in Medical Applications ... 179
Narges Manouchehri, Maryam Rahmanpour, and Nizar Bouguila
9.1 Introduction ... 180
9.2 Finite Inverted Dirichlet Mixture Model ... 182
9.3 Entropy-Based Variational Learning ... 183
9.3.1 Variational Learning ... 183
9.3.2 Model Learning Through Entropy-Based Variational Bayes ... 186
9.3.3 Theoretical Entropy of Inverted Dirichlet Mixtures ... 186
9.3.4 MeanNN Entropy Estimator ... 187
9.4 Experimental Results ... 188
9.4.1 Cardiovascular Diseases (CVDs) ... 189
9.4.2 Diabetes ... 190
9.4.3 Lung Cancer ... 191
9.4.4 Breast Cancer ... 192
9.5 Conclusion ... 193
References ... 193

Contributors

Oussama Barakat Nanomedecine LAB, University of Bourgogne Franche-Comté, Besançon, France
Faculté of Sciences and Technologies, Besançon, France

Mouna Berquedich Laboratory of Innovative Technologies (LTI), Abdelmalek Saâdi University, Tangier, Morocco
Innovation Lab for Operations (ILO), University Mohammed 6 Polytechnic, Ben Guerir, Morocco

Henri J. Boersma Maastricht University Medical Centre, Maastricht, Netherlands

Nizar Bouguila Concordia Institute for Information Systems Engineering, Concordia University, Montreal, QC, Canada

Laurent Deshayes Innovation Lab for Operations (ILO), University Mohammed 6 Polytechnic, Ben Guerir, Morocco

Souhir Elleuch Department of Management Information Systems, College of Business and Economics, Qassim University, Buraidah, Qassim, Saudi Arabia

Khalid M. Hosny Faculty of Computers and Information, Zagazig University, Zagazig, Egypt

Bassem Jarboui Higher Colleges of Technology, Abu Dhabi, UAE

Meeta Kalra Concordia Institute for Information Systems Engineering, Concordia University, Montreal, QC, Canada

Oulaid Kamach Laboratory of Innovative Technologies (LTI), Abdelmalek Saâdi University, Tangier, Morocco

Asmaa M. Khalid Faculty of Computers and Information, Zagazig University, Zagazig, Egypt

Niklas Kühl Karlsruhe Service Research Institute (KSRI), Karlsruhe Institute of Technology (KIT), Karlsruhe, Germany

Narges Manouchehri Concordia Institute for Information Systems Engineering (CIISE), Concordia University, Montreal, QC, Canada

Malek Masmoudi Faculty of Sciences and Technologies, University of Jean-Monnet, Saint-Étienne, France

Taïeb Mellouli Department of Business Information Systems and Operation Research, Martin-Luther-University Halle-Wittenberg, Halle, Germany

Nenad Mladenovic Research Center of Digital Supply Chain and Operations, Department of Industrial and Systems Engineering, Khalifa University, Abu Dhabi, UAE

Jyoti R. Munavalli BNM Institute of Technology (VTU), Maastricht University, Bangalore, India

Maryam Rahmanpour Concordia Institute for Information Systems Engineering (CIISE), Montreal, QC, Canada

Shyam Vasudeva Rao Maastricht University Medical Centre, Maastricht, Netherlands

Melanie Reuter-Oppermann Karlsruhe Service Research Institute (KSRI), Karlsruhe Institute of Technology (KIT), Karlsruhe, Germany

Olga Rusetskaya ICSER Leontief Centre, St. Petersburg, Russia

Ehab R. Mohamed Faculty of Computers and Information, Zagazig University, Zagazig, Egypt

Bilal Bou Saleh University of Bourgogne Franche Comté, Besançon, France
Lebanese University, Beirut, Lebanon

Ghazi Bou Saleh Lebanese University Faculty of Technology, Saida, Lebanon

Thomas Stoeck Department of Business Information Systems and Operation Research, Martin-Luther-University Halle-Wittenberg, Halle, Germany

G. G. van Merode Maastricht University Medical Centre, Maastricht, Netherlands

Acronyms

AI	Artificial intelligence
AIRS	Artificial immune recognition system
AIS	Artificial immune system
AMI	Adjusted mutual information
ANN	Artificial system and neural network
AP	Automatic programming
AR	Agent room
ARI	Adjusted Rand index
BV-DMM	Batch variational learning of Dirichlet mixture models
BV-GMM	Batch variational learning of Gaussian mixture models
BV-IDMM	Batch variational learning of inverted Dirichlet mixture models
CA	Cluster analysis
CAD	Computer-aided detection
CHU	Centre hospitalier universitaire
CI	Computational Intelligence
CNN	Convolution neural network
CNP	Contract net protocol
CR	Compression ratio
CT	Computed tomography
CVDs	Cardiovascular diseases
DA	Department agent
DAI	Distributed artificial intelligence
DCT	Discrete cosine transform
DICOM	Digital imaging and communications in medicine
ED	Emergency department
EMS	Emergency medical services
EV-DMM	Entropy-based variational learning of Dirichlet mixture models
EV-IDMM	Entropy-based variational learning of inverted Dirichlet mixture models
E. coli	Escherichia coli
EHR	Electronic health records

EID	Electronic health information exchange
FM	Fowlkes–Mallows index
FSVM	Fuzzy support vector machine
GBM	Gradient boosted machines
G-DRG	German Diagnosis Related Groups
GD	Generalized Dirichlet
GDP	Gross domestic product
GHT	Hospital Group Territory
GID	Generalized inverted Dirichlet
GMM	Gaussian mixture model
GP	Genetic programming
HIT	Health information technology
HMO	Health maintenance organizations
HR	Human resources
ICU	Intensive care units
Id	Identification
ID	Inverted Dirichlet
IDMM	Inverted Dirichlet mixture model
IMA	Immune memory algorithm
IoT	Internet of things
IT	Information technology
KL	Kullback–Leibler
KPI	Key performance indicator
LMs	Legendre moments
MA	Manager agent
MAP	Multi-agent planer
MAS	Multi-agent system
MCMC	Markov chain Monte Carlo
MILP	Mixed integer linear programming
ML	Machine learning
MRI	Magnetic resonance imaging
MSE	Mean square error
NCC	Normalized correlation coefficient
NEMA	The National Electrical Manufacturers Association
NF	Neuro-fuzzy
NSA	Negative selection algorithm
OECD	Organisation for Economic Co-operation and Development
OPC	Out-patient clinic
OR	Operating room
OVGIDMM	Online variational learning of finite GID with feature selection
OVGMM	Online variational learning of finite Gaussian mixture model
OVIDMM	Online variational learning of finite inverted Dirichlet mixture model
PA	Patient agent
PFSA	Probabilistic finite state automata
PSA	Patient scheduling agent

PSNR	Peak signal-to-noise ratio
PSO	Particle swarm optimization
RA	Resource agent
RNC	Required number of coefficients
ROA	Route agent
ROI	Region of interest
RSA	Resource scheduler agent
SARIMA	Seasonal autoregressive integrated moving average
SPIHT	Set partitioning in hierarchical trees
SSIM	Structural similarity index measure
SVM	Support vector machine
SYR	Symbolic regression
TCIA	The Cancer Imaging Archive
VNP	Variable neighborhood programming
VNS	Variable neighborhood search
WHO	World Health Organization
WOA	Whale optimization algorithm

Chapter 1
Artificial Intelligence for Healthcare Logistics: An Overview and Research Agenda

Melanie Reuter-Oppermann and Niklas Kühl

Abstract In this chapter we present the existing literature on machine learning approaches and artificial intelligence for logistical problems arising for designing, providing and improving healthcare services. As a basis, we provide a framework for the classification of artificial intelligence. For the analysis, we distinguish between the care levels (primary, secondary and tertiary care), the planning levels (strategic, tactical and operational), as well as the user types (doctors, nurses, technicians, patients, etc.). Based on the results, we provide a research agenda with open topics and future challenges.

1.1 Introduction

The techniques of *machine learning* (ML) and *artificial intelligence* (AI) are omnipresent in today's academic discussions [1]. In this chapter, we aim to shed light on the capabilities of artificial intelligence for the area of healthcare logistics, a promising field in operations research [2]. Based on a literature review [3], we explore three different aspects of interest to reveal existing research as well as future possibilities. First, an overview of the care levels [4], i.e. *primary*, *secondary* and *tertiary* care and existing as well as future AI applications in this dimension requires analysis. Second, the aspect of planning—distinguished into *strategic*, *tactical* and *operational* levels—is of interest [5]. Finally, we regard the user types of the healthcare logistic services [6], e.g. *doctors*, *technicians* or *patients* and possible enhancements of their tasks with AI. Therefore, we contribute to the body

M. Reuter-Oppermann (✉)
Information Systems, Software & Digital Business Group, Technical University of Darmstadt, Darmstadt, Germany
e-mail: oppermann@is.tu-darmstadt.de; melanie.reuter@kit.edu

N. Kühl
Karlsruhe Service Research Institute (KSRI), Karlsruhe Institute of Technology (KIT), Karlsruhe, Germany
e-mail: niklas.kuehl@kit.edu

M. Masmoudi et al. (eds.), *Artificial Intelligence and Data Mining in Healthcare*,
https://doi.org/10.1007/978-3-030-45240-7_1

of knowledge by providing a holistic overview of AI in healthcare logistics and derive a research agenda to highlight priorities in future research endeavours in this highly important field.

This work focuses only on approaches applied to healthcare logistics, i.e. planning problems arising for logistical tasks, e.g. in hospitals or other care institutions. That is, AI for medical applications, e.g. to determine the probability for a certain disease, is excluded from this work.

The remainder of this work is structured as follows: we first set the required nomenclature as a basis by summarising the concepts of machine learning and artificial intelligence in Sect. 1.2. With the necessary definitory framework at hand, we review the aspects of care levels, the planning levels as well as the user types in Sect. 1.3. We then categorise the existing literature in Sect. 1.4, synthesise our findings within a holistic overview and derive a research agenda accordingly. We conclude with recommendations, a summary and limitations in Sect. 1.5.

1.2 Machine Learning and Artificial Intelligence

As a basis for our classification of related work and the resulting research agenda, we first review the different notions and concepts of machine learning and artificial intelligence within extant literature. In addition, we come up with a working definition of artificial intelligence for the remainder of this work.

Machine learning and artificial intelligence are related, often present in the same context and sometimes used interchangeably [7]. While the terms are common in different communities, their particular usage and meaning vary widely—especially with the rise of AI research within the past decade [8].

1.2.1 Machine Learning

Within the field of computer science, machine learning has the focus of designing efficient algorithms to solve problems with computational resources [9]. While machine learning utilises approaches from the field of statistical learning [10], it also includes methods that are not entirely based on previous work of statisticians—resulting in new and well-cited contributions to the field [11–13]. Recently, especially the method of deep learning raised increased interest [14], as it has drastically improved the capabilities of machine learning, e.g. in speech [15] or image recognition [16].

In its basic form, machine learning describes a set of techniques that are used to solve a variety of real-world problems with the help of computer systems that can learn to solve a problem instead of being explicitly programmed [17]. In general, we can differentiate between supervised, unsupervised and reinforcement learning [18, 19]. *Supervised learning* comprises methods and algorithms to learn the mapping from the input to the output. For instance, letting a child sort toy cars and telling the child in advance there are sports cars and SUVs, the child would perform a supervised learning task. It would learn to recognise patterns from the regarded cars (input) and sort them accordingly (output). *Unsupervised learning*, however, comprises methods and algorithms that are able to reveal previously unknown patterns in data. In the example discussed before, letting a child sort toy cars and letting the child determine how to arrange/cluster them would be an unsupervised learning task. In demarcation from supervised learning, *reinforcement learning* differs in the fact that correct input/output combinations need not be presented, and sub-optimal actions need not be explicitly corrected. Instead the focus is on finding a balance between the exploration of uncharted solutions and the exploitation of already achieved knowledge [20]. In case of the child sorting toys, this would mean he or she would (sometimes) receive feedback ("rewards") on the nature of its decisions, which allows it to slowly build up additional knowledge. In the field of operations research, reinforcement learning is also called *approximate dynamic programming* [21], or *neuro-dynamic programming* [22].

1.2.2 Artificial Intelligence

The topic of artificial intelligence (AI) is rooted in different research disciplines, such as computer science, philosophy, or futures studies [7]. In this work, we mainly focus on the field of computer science, as it is the most relevant one in identifying the contribution of AI to the field of healthcare logistics. AI research can be separated into different research streams [23]. These streams differ on the one hand as to the objective of AI application—thinking vs. acting, and on the other hand, as to the kind of decision-making—targeting a human-like decision vs. an ideal, rational decision. Especially the research stream of "Rational Agents" considers an AI as a rational [23] or intelligent [24] agent. This stream is the most relevant for the remainder of this work, as it describes the implementation of AI as intelligent agents within real-world environments.

Machine learning plays three major roles in this field of artificial intelligence. First, machine learning is relevant in the implementation of intelligent agents, precisely in the back-end layer of such agents. When regarding the case of supervised and reinforcement machine learning, we need to further differentiate between the process task that is building (equivalently training) adequate machine learning models [25] and the process task that is executing on the knowledge from these models [26]. Therefore, the "thinking layer" of agents is typically defined by machine learning models [7].

Second, the learning back-end in the intelligent agent dictates if and how the agent is able to learn, e.g. which precise algorithms it uses, what type of data processing is applied, how concept drift [27] is handled, etc. Russel and Norvig [23] regard two different types of intelligent agents: *simple-reflex agents* and *learning agents*. This differentiation considers whether the underlying models in the thinking layer are once trained and never touched again ("simple-reflex agent") or continuously updated and adaptive ("learning agent"). In the recent literature, suitable examples for both can be found [28–30].

Third, it is of interest how automated the necessary process steps are for an AI. Every machine learning task involves various process steps, including data source selection, data collection, pre-processing, model building, evaluating, deploying, executing and improving [31]. The autonomy and the automation of these tasks are of particular interest, especially the necessary human involvement for the AI to execute [7].

1.2.3 Working Definition

Synthesising the results from the previous two sections, we depict our definitory framework of AI in Fig. 1.1. We differentiate on the different process tasks of data analysis, recommendation and decision/action for any AI endeavour. *Data analysis* includes all necessary steps of pre-processing and automated generation of predictions based on new, incoming data—typically based on machine learning. The task of *recommendation* describes the interpretation of the results from the previous task. For instance, an AI might just deliver outputs on whether or not a certain ambulance service will make it in time to the hospital. The more advanced step would be to output recommendations, e.g. using a different available ambulance service that could arrive more promptly. Finally, the last process task is making a *decision* and possibly *act* on it. In the previous example, this would mean the dispatching of a different ambulance. We differentiate which of these tasks the AI handles and which is handled by humans, resulting in the three levels of *AI-Enrichment*, *AI-Enhancement* and *AI-Autonomy*.

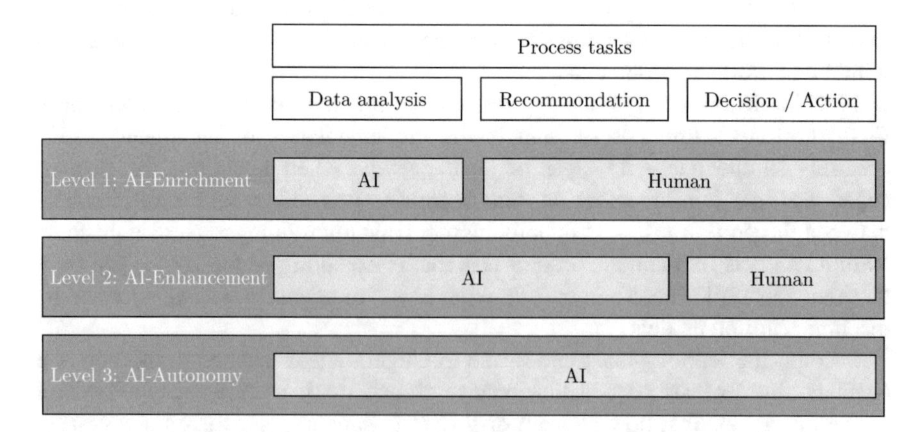

Fig. 1.1 Levels of AI involvement for different process tasks

1.3 Framework for Healthcare Logistics Literature

The review of prior literature is an integral part of any research project, as a comprehensive overview allows the necessary foundations for advancing knowledge. Furthermore, it allows to uncover research gaps, where additional work is needed. According to Webster and Watson [3], one of the most used literature review methods [32], each review of the existing body of knowledge is separated into multiple steps.

The first step deals with the collection of articles. The collected articles must reflect the research topic in its entirety. The search should not be limited to a specific journal, geographical area or research method. However, one typically starts to search for relevant articles in leading specialist journals. These then serve as a starting point for a *forward–backward search*. The *backward search* is the search for further relevant literature from the cited sources of the articles in question. The *forward search* is the search for the literature that quotes the present article by searching the article, e.g. via the "Web of Science".

In the second step, the selected literature is analysed. This is done concept—and not author—related, because the review article should be based on ideas and concepts and not on people. The identification of research gaps, the lack of essential concepts in the review and the use of the article as a summary of the existing literature for further research articles are thus facilitated. A concept matrix facilitates the sorting of the relevant articles by categorising the articles according to defined variables, analysis levels, etc.

Third, Webster and Watson recommend the elaboration of gaps in research. They also recognise that this is the most difficult part of a literature analysis. A concept model with supporting propositions is considered to be the methodology for the development of this task. They point out that this model and its propositions gain importance only through the justification of the relationship between the chosen variables.

Finally, the last step deals with the evaluation, conclusion and discussion of the obtained results.

In the following, a framework for classifying the published literature is developed. It bases on three dimensions: planning levels, care levels and user types. These three aspects are first defined separately and then combined into one framework.

1.3.1 Planning Levels

Healthcare logistics can be divided into the three planning levels: strategic, tactical and operational (Fig. 1.2). While strategic decisions are usually made for years or even decades, tactical decisions can be revoked on a yearly or monthly basis. Operational planning happens daily, often as offline decisions, or in real-time usually with online approaches.

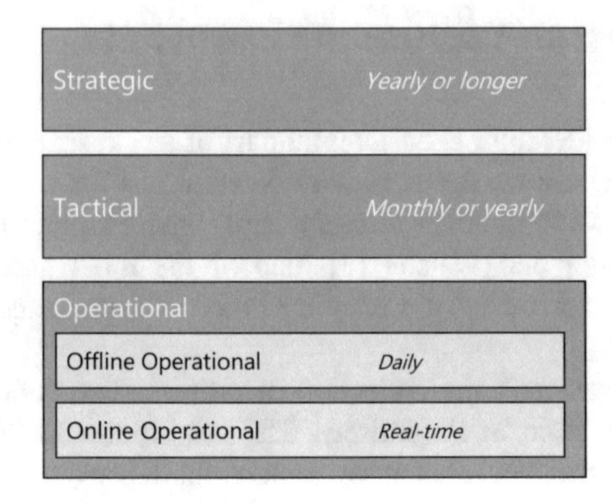

Fig. 1.2 Planning levels in healthcare logistics

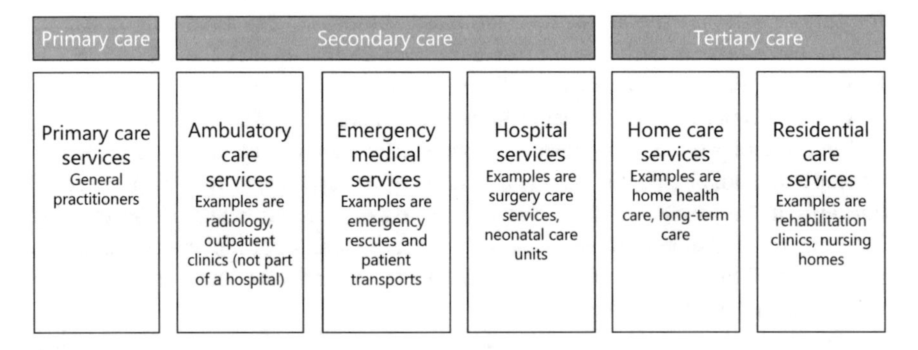

Fig. 1.3 Levels of medical care

1.3.2 Care Levels

In general, we differentiate between primary, secondary and tertiary care (Fig. 1.3). Secondary care includes ambulatory care services, emergency medical services and hospital services. Tertiary care consists of home care services and residential care services. Hulshof et al. [33] have proposed a similar structure and used it to categorise existing research for planning problems in healthcare for which operations research literature has been published.

1.3.3 User Types

In healthcare logistics, users of AI approaches could be manifold due to the nature and design of the underlying service network and its contributors and participants. This includes multiple stakeholders, precisely:

- the patient who receives treatment,
- insurance companies who might pay for the service,

- relatives who desire to be informed and might do part of the care at home,
- hospital staff, including
  - managers,
  - doctors (general practitioners and specialists),
  - nurses,
  - medical information scientists,
  - laboratory staff,
  - physiotherapists,
  - transport services,
  - IT.
- emergency medical services, including
  - dispatchers,
  - paramedics,
  - drivers,
  - relief organisations,
- as well as the government (e.g. health ministries and other policy-makers).

A framework would not be useful if all possible user types were included. Also, not all users have been targeted in the literature. Therefore, we distinguish the following three user types in our framework:

- Patients: Individuals who receive medical care from providers.
- Providers: Institutions or people that provide care to patients and charge payers for that care. We divide this type again into two subgroups:
  - Hospital management, and
  - doctors and nurses.
- Payers: Institutions that pay providers for healthcare services, which include insurance carriers, private employers and the government.

1.3.4 Framework

Figure 1.4 shows the framework that we will use to structure the literature on AI applied to healthcare logistics that is based on the one introduced by Hulshof et al. [33]. The overall four user types are abbreviated by P1–P3, with P1 meaning patients, P2.1 hospital management, P2.2 doctors and nurses, and P3 payers. References are included with their respective number [X].

	Primary care	Secondary care			Tertiary care	
	Primary care services	Ambulatory care services	Emergency medical services	Hospital services	Home care services	Residential care services
Strategic	P1 [X1]					
Tactical		P2.1 [X2.1]				
Operational						
Offline Operational			P2.2 [X2.2]			
Online Operational				P3 [X3]		

Fig. 1.4 Framework for sorting the literature on AI in healthcare logistics

1.4 Literature Review

As described in the previous section, we performed a literature analysis with Google Scholar, using the search strings as shown in Fig. 1.5.

When scanning the literature, we classified papers on AI for healthcare in five main categories:

1. Publications that apply AI to medical problems, e.g. to determine the probability for a certain disease, the probability that a certain treatment will be successful or survival predictions, e.g. in the intensive care unit (ICU) [34, 35]. Examples of commonly studied areas include stroke prediction [36], infections, diabetes, e.g. personalisation of diabetes therapy [37] or predicting hospital readmissions for diabetics [38], and cancer. Machine learning approaches can predict rehabilitation potential [39]. Researched topics also include AI-based decision support systems for personalised medicine and treatment [40] as well as trends in telemedicine with AI [41]. Neural networks can be used to predict future illnesses that can be of interest for insurance companies to determine expected costs [42]. In addition, several AI approaches have been proposed for

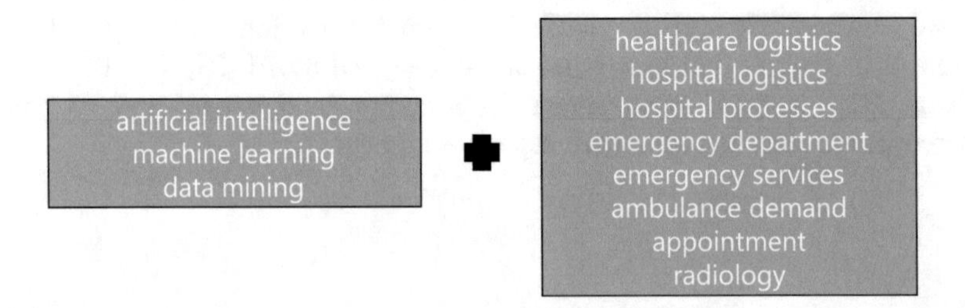

Fig. 1.5 Search strings used for the literature review

radiotherapy. For example, AI can be applied for radiology [43] or radiotherapy treatments to determine patient-specific dosing [44]. The majority of publications are patient-centred with a decision support focus on doctors.
2. Especially in the USA, AI approaches are applied to the scanning of medical claims and the detection of fraud. These approaches are especially relevant for insurance companies and hospital management.
3. Several review papers within the interplay of analytical techniques and healthcare have been published—each focusing on a different topic, e.g. AI capabilities [45, 46], data mining [47–53], big data in general [54], business intelligence [55], as well as a systematic mapping study [56], healthcare organisational decision-making [57] and specific reviews on big data and machine learning in radiation oncology [58] or artificial intelligence for neonatal care [59].
4. One of the two relevant categories for this work are publications in which AI methods are used to predict input for healthcare logistics problems, e.g. predicting demand for healthcare services like ambulances [60], surgery durations or no-shows [61].
5. In the last category, publications embed AI into optimisation approaches for healthcare logistics problems, e.g. as part of a heuristic.

In our search, we identified 132 papers to be relevant for this work based on the title and the keywords. After having read the papers, we discarded 81 papers. We derived two relevant main categories: AI for optimisation input (34 papers) as well as AI for healthcare logistics optimisation (17 papers)—both of which will be regarded in detail in the following.

1.4.1 AI for Optimisation Input

Menke et al. present an artificial neural network (ANN) to predict the patient volume arriving at an emergency department (ED) [62]. The authors state that the ANN can be used for this task, but it must be properly designed and include all relevant variables. Then, it can be used as an input for staffing the ED appropriately, aiming at shorter waiting times for patients as well as lower or more balanced workloads for the doctors and nurses. Afilal et al. propose time series-based forecasting models to predict long-term as well as short-term arrival rates at the emergency department [63]. Also Xu et al. target the prediction of daily arrivals at the ED [64], especially considering non-critical patients. They investigate an artificial neural network (ANN) to study the contributing variables and daily arrival rates. They compare it with the non-linear least square regression (NLLSR) and the multiple linear regression (MLR). Acid et al. use different algorithms for learning Bayesian networks to predict arrivals at the ED [65].

Besides the overall patient volume, predicting the admission rates is another important problem arising for emergency departments that can be addressed by machine learning approaches. Graham et al. compare three machine learning approaches: logistic regression, decision trees and gradient-boosted machines (GBM) for predicting hospital admissions from the ED [66]. The authors find that GBM leads to the highest accuracy, while logistic regression should be chosen when interpretability is most important. Having good estimates for admission rates can improve patient flow and help schedule resources. Krämer et al. use supervised machine learning techniques to classify hospital admissions into urgent and elective care [67]. Patient disposition decisions are predicted by Lee et al. [68]. They distinguish four admission classes: intensive care unit, telemetry unit, general practice unit and observation unit, and test three different machine learning approaches for the classification.

Length of stay (LOS) is one of the main ED performance indicators. Benbelkacem et al. test and compare several machine learning approaches for predicting a patient's LOS, for example based on date and hour [69].

Cai et al. choose a Bayesian network to develop a model for real-time predictions of length of stay, mortality and re-admission for hospitalised patients based on ED data [70].

In order to prioritise patients' admission to the hospital, Luo et al. test logistic regression, random forest, gradient-boosting decision tree, extreme gradient boosting and combination of all four approaches [71].

For a hospital, patient re-admissions are a big challenge. Machine learning methods can help detect patients with a high re-admission probability. Turgeman and May combine a boosted C5.0 tree, as the base classifier, with a support vector machine (SVM), as a secondary classifier [72]. Zheng et al. study classification models that use neural networks, random forest and support vector machines for hospital re-admissions, with a specific focus on heart failure patients [73]. A clinical tool based on machine learning for predicting patients who will return within 72 h to the paediatric emergency department was developed by Lee et al. [74]. Futoma et al. compare predictive and deep learning models for 30 day re-admissions [75]. The re-admission problem is also studied by Rana et al. [76], Eigner et al. [77] and Wickramasinghe et al. [78].

Beds in ICUs are often scarce resources. Then, it is important to timely discharge patients to a regular ward when possible. McWilliams et al. test two machine learning classifiers, a random forest and a logistic classifier, to detect patients who can be discharged from the ICU [79].

Padoy envisions a decision support tool for workflow recognition during surgery using machine learning approaches, more specifically deep learning [80]. He proposes to use video input, e.g. laparoscopic videos, to automatically determine the current phase of the surgery.

Funkner et al. use a decision tree approach for predicting clinical pathways in a hospital [81].

Al Nuaimi presents four models based on naive Bayes, k-nearest neighbour, and SVM to predict the healthcare demand in Abu Dhabi, i.e. a district's current and future needs for hospitals and clinics [82].

No-shows play an important role for appointment planning, in hospitals as well as private practices. Topuz et al. present a Bayesian belief network for predicting paediatric clinic no-shows [83]. Nelson et al. try to predict no-shows for scheduled magnetic resonance imaging appointments using logistic regression, support vector machines, random forests, AdaBoost and gradient-boosting machines [84]. Goffman et al. apply logistic regression models to predict no-shows and appointment behaviour in the Veterans Health Administration [85]. Harris et al. combine regression-like modelling and functional approximation, using the sum of exponential functions, to predict patient no-shows [86]. A hybrid probabilistic model based on multinomial logistic regression and Bayesian inference to predict the probability of no-show and cancellation in real time was developed by Alaeddini et al. [87]. Goldman et al. apply a multivariate logistic regression analysis to predict the no-show rate in primary care [88]. Kurasawa et al. predict no-shows for clinical appointments by diabetic patients [61].

Due to high-cost pressures in many healthcare systems, it is crucial for hospitals to avoid unnecessary high spending. Accordingly, Eigner et al. propose to use machine learning approaches to identify high-cost patients [89].

Outside of hospitals, several publications have used machine learning approaches to predict the demand for emergency medical services (EMS), i.e. ambulances. Good predictions can be used as input for ambulance location and relocation approaches to reduce response times. Spatio-temporal predictions for 1-h intervals and 1 km^2 with three methods based on Gaussian mixture models, kernel density estimation and kernel warping are proposed by Zhou [90]. Chen and Lu test moving average, artificial neural network, linear regression and support vector machine to predict emergency services demand [91]. Chen et al. apply moving average, artificial neural network, sinusoidal regression and support vector regression to the same problem [60]. Villani et al. focus on forecasting pre-hospital diabetic emergencies that have to be served by an Australian EMS provider, using the seasonal autoregressive integrated moving average (SARIMA) modelling process [92].

Curtis et al. studied the applicability of machine learning models to predict waiting times at a walk-in radiology facility as well as waiting times for patients with scheduled appointments at radiology facilities [93]. The authors test and compare many different machine learning algorithms, including neural network, random forest and support vector machine.

In Fig. 1.6, the publications are inserted into the framework.

1.4.2 AI for Healthcare Logistics Optimisation

Arnolds and Gartner propose a machine learning-based clinical pathway mining approach for hospital layout planning [94]. The machine learning algorithm builds on the probabilistic finite state automata (PFSA) to learn significant clinical pathways. The pathways are then one input for the mathematical model to determine a hospital layout. In their case study, the determined layout reduces distances travelled by patients significantly.

	Primary care	Secondary care			Tertiary care	
	Primary care services	Ambulatory care services	Emergency medical services	Hospital services	Home care services	Residential care services
Strategic			P2 [60], [90], [91], [92]	P2.1 [81], [82]		
Tactical	P2.2 [88]	P2.2 [83], [84], [85], [86], [87]		P2.1 [61], [66] [89]		
Operational						
Offline Operational				P2.1/P2.2 [62], [63], [64], [65]		
Online Operational		P2.2 [93]		P2.2 [67], [68], [69], [70],...,[80]		

Fig. 1.6 Overview of publications on AI for optimisation input

Gartner et al. use ML methods for predicting and classifying diagnosis-related groups [95]. They incorporate the methods into an MIP-based resource allocation model. They showed that their approach could improve diagnosis-related group (DRG) classification both before and during the admission of a patient. The optimised allocation of resources such as operating rooms and beds allowed an additional 9% of elective patients to be admitted.

Alapont et al. discuss the design and the use of a data mining tool for hospital management, based on algorithms that are implemented in Weka [96]. They connect logistical problems like bed management, surgery scheduling and resource allocation to machine learning-based predictions of bed occupation, hospital admission rates and emergencies, for example.

Also Srikanth and Arivazhagan propose to use machine learning approaches for demand prediction in order to improve hospital resource allocations and patient schedules [97]. They present a patient inflow prediction model based on a resilient back-propagation neural network, which improves the prediction accuracy.

Again with the aim of improving the allocation of resources in a hospital, Aktaş et al. developed a decision support system that contained a Bayesian belief network [98].

In order to improve processes in an emergency department, Laskowski combines an agent-based model with a machine learning approach to a hybrid system for decision support [99].

Ceglowski et al. combine a data-mining approach with a discrete-event simulation in order to analyse processes in an emergency department and to identify bottlenecks in the interface between the ED and the hospital wards [100].

Thompson et al. summarise and discuss artificial intelligence for radiation oncology [101]. Besides clinical decision support, image segmentation and dose optimisation, they also mention the importance of and potential gains for logistical problems like staffing and resource allocation.

Lofti and Torres present a predictive model to be used in scheduling patients in an urban outpatient clinic [102]. They use decision tree methods to determine the likelihood of a patient's no-show and test the results in a scheduling algorithm.

Patient no-shows also play an important role in the work of Srinivas and Ravindran [103]. The authors developed a prescriptive analytics framework to schedule patient appointments for a family medicine clinic. Machine learning was used to classify patients based on their no-show risk. Improving patient satisfaction was one of the main aims.

Harper presents a framework for machine learning approaches like decision trees with healthcare optimisation models to improve healthcare processes and resource utilisation [104].

An ambulance location approach using predicted future emergency locations by an artificial neural network was published by Grekousis and Liu [105]. In order to do so, they introduced new concepts and notions to model emergency events as sets of interconnected points in space that create paths over time.

Two further publications do not explicitly match the focus of this chapter but present very interesting approaches. The first uses data mining and machine learning in the context of disaster and crisis management [106]. The authors state that the approaches allow to address a wider spectrum of problems, such as situational awareness and real-time threat assessment using diverse streams of data. The second presents an ML-based approach for automatic stress detection in emergency phone calls to support dispatching of ambulances and help call takers distinguish important and critical calls in case of high workloads [107].

In addition to the previously discussed papers, we found three additional, rather recent, works that present promising optimisation approaches with embedded machine learning methods.

Potter develops two machine learning-based greedy heuristics to improve the assignment of patients to the available treatment machines in radiotherapy [108]. Strahl aims at reducing waiting times and surgeon idle time in an outpatient clinic [109]. He uses supervised learning regression to predict appointment durations and surgeon arrival times in a scheduling approach, a greedy hill-climbing algorithm, to improve the existing schedule. Letzner proposes supervised machine learning approaches for predicting hospital selection when a patient is picked up by an ambulance [110]. The performance of random forest, logistic regression and neural network are compared.

In Fig. 1.7, the publications are inserted into the framework.

1.4.3 *AI for ED Logistics*

We now want to take a closer look at one exemplary area in healthcare logistics for using AI. The overview shows that comparably many publications target

	Primary care	Secondary care			Tertiary care	
	Primary care services	Ambulatory care services	Emergency medical services	Hospital services	Home care services	Residential care services
Strategic			P2 [105]	P2.1 [94]		
Tactical	P2.2 [103]	P2.2 [101], [102]		P2.1 [95], [96], [97], [98], [104] P2.2 [99], [100]		
Operational						
Offline Operational		P2.2 [108], [109]				
Online Operational			P2 [107], [110]			

Fig. 1.7 Overview of publications on AI for healthcare logistics optimisation

emergency departments. Crowding is becoming a challenging issue for many hospitals worldwide. The main uncertainties for ED and hospital managers are the arrival rates of patients, together with their necessary treatments and treatment times, as well as the probabilities for patients being admitted to the hospital. Out of the 12 ED-related papers in this review, three predict arrival rates or ED volume [62–64], and one predicts re-arrivals at the ED [74], two consider patients' admission from ED [66, 67], one studies more concretely disposition from ED [68], two target LOS [69, 70] and three more generally ED management [65, 99, 100]. Eight papers only use a single machine learning approach, altogether four papers use two [64], three [66] or four [68, 69] different approaches and compare the results. The 12 papers apply 14 different machine learning approaches altogether, whereas artificial neural networks and logistic regressions are used in three publications, support vector machines and decision trees are studied twice. Random forest, for example, is used in one paper. The variety of approaches used makes it difficult to draw general conclusions on which approach to use for which problem. Future research could investigate this question and study methods in detail, preferably using different data sets, as data items used in the publications also differ significantly. While most use patient arrival times and dates, some use patient-related data like age and gender or case-related data items like the triage score and treatment durations, while others also use external data items like weather. Overall, we counted more than 25 input parameters.

1.4.4 Synthesis and Research Agenda

Accurate predictions are basically important for all planning problems in healthcare for which the input is not completely deterministic—which is hardly ever the case in reality. This is true for all care levels, planning levels and user types. The prediction types and levels that are necessary at different planning levels vary. While it might be sufficient for an emergency services provider to know the average daily occurrence of emergencies in the next years when locating the bases at the strategic level, on the operational level a precise prediction for ambulance demand in the next hour is of interest. This often leads to the fact that different (machine learning) methods must be tested and deployed for predicting demand, service times or other inputs at the three planning levels. Predicting demand at one level might be easier than at another.

Most of the publications included in this study focus on one care level and one planning level, while it would be interesting and promising to test the approaches also for other care and planning levels. Then, many of the "gaps" in Figs. 1.6 and 1.7 can be easily filled. Resulting research questions for machine learning and artificial intelligence should also address the transferability of approaches between providers.

Future research has the potential to address previously untapped potential in the application of AI for healthcare management and the optimisation of healthcare logistics, including medical predictions and personalised treatments. As a next step, the question arises how predictions can be incorporated best into existing processes like scheduling an assignment of resources—as so far, research in this area is still rare.

Most of the approaches have only been presented and tested in theory so far. Hardly any were actually included in decision support systems and deployed in practice. Arising research questions include (1) *How well do the predictions perform in practice?* and (2) *What is the quality of the optimisation approaches that are based on or make use of AI?*

1.5 Conclusion

In this chapter we have presented a framework for classifying the literature on AI applied to healthcare logistics. The framework integrates care levels, planning levels and user types.

Figures 1.6 and 1.7 show the publications inserted into the framework.

While the majority of papers we came across used machine learning approaches for medical purposes, more and more publications address healthcare logistics and management problems. The literature overview shows that applying artificial intelligence and especially machine learning methods to healthcare logistics problems is a promising approach. A few areas have already been addressed by several publications, including appointment planning, (patient) scheduling and resource utilisation. Machine learning methods have been successfully developed to predict demand for emergency departments, the ICU or ambulances, for example.

Certain limitations apply to the presented research. The quality of our systematic literature review is assessed based on a set of quality assessment criteria as introduced by Kitchenham et al. [111]. Inclusion and exclusion criteria are described in detail in Sect. 1.4. In terms of the coverage of our study, we used all literature that Google Scholar provided us for our search terms. Regarding the assessment of the quality and validity of the analysed literature, we did not assess the validity of the studies, for example by journal rankings or citations, as our aim was of exploratory nature. Furthermore, the extracted knowledge from scientific literature is always liable to the specific scientific environment, e.g. there can be biases in the methods that were used [112].

Another challenge we faced when searching the literature was a missing clear definition of machine learning and artificial intelligence methods. While authors always named the method they applied, not all papers used the terms "machine learning" or "artificial intelligence", making it harder to find and to classify them. In this work, we included all papers that either used at least one of the two terms or applied a method that can be categorised as AI by common understanding as well as our own introduced understanding of the terms.

Based on the literature overview, three areas for future research were detected. The first is studying the transferability of existing approaches to other care and planning levels as well as to providers on the same levels. The second addresses the need to further include machine learning and artificial intelligence approaches into healthcare management and optimisation of planning problems, also making use of already existing AI methods. The third area regards the actual deployment of AI methods in practices, potentially integrated into decision support systems.

References

1. Alok Aggarwal. The current hype cycle in artificial intelligence. *Retrieved from KDNuggets*, 2018.
2. Sylvain Landry and Richard Philippe. How logistics can service healthcare. In *Supply Chain Forum: An International Journal*, volume 5, pages 24–30. Taylor & Francis, 2004.
3. Jane Webster and Richard T Watson. Analyzing the past to prepare for the future: Writing a literature review. *MIS quarterly*, pages xiii–xxiii, 2002.
4. Oliver Gröne and Mila Garcia-Barbero. Integrated care. *International journal of integrated care*, 1(2), 2001.
5. Erwin W Hans, Mark Van Houdenhoven, and Peter JH Hulshof. A framework for healthcare planning and control. In *Handbook of healthcare system scheduling*, pages 303–320. Springer, 2012.
6. Robert G Fichman, Rajiv Kohli, and Ranjani Krishnan. Editorial overview the role of information systems in healthcare: current research and future trends. *Information Systems Research*, 22(3):419–428, 2011.

7. Niklas Kühl, Marc Goutier, Robin Hirt, and Gerhard Satzger. Machine learning in artificial intelligence: Towards a common understanding. In *Proceedings Hawaii International Conference on Systems Sciences*, 2019.
8. Hidemichi Fujii and Shunsuke Managi. Trends and priority shifts in artificial intelligence technology invention: A global patent analysis. *Economic Analysis and Policy*, 58:60–69, 2018.
9. Mehryar Mohri, Afshin Rostamizadeh, and Ameet Talwalkar. *Foundations of machine learning*. MIT press, 2012.
10. Olivier Bousquet, Ulrike von Luxburg, and Gunnar Rätsch. *Advanced Lectures on Machine Learning: ML Summer Schools 2003, Canberra, Australia, February 2–14, 2003, Tübingen, Germany, August 4–16, 2003, Revised Lectures*, volume 3176. Springer, 2011.
11. Guang-Bin Huang, Qin-Yu Zhu, and Chee-Kheong Siew. Extreme learning machine: a new learning scheme of feedforward neural networks. In *Neural Networks, 2004. Proceedings. 2004 IEEE International Joint Conference on*, volume 2, pages 985–990. IEEE, 2004.
12. Fabrizio Sebastiani. Machine learning in automated text categorization. *ACM Computing Surveys*, 34(1):1–47, 2002.
13. Ian Goodfellow, Jean Pouget-Abadie, Mehdi Mirza, Bing Xu, David Warde-Farley, Sherjil Ozair, Aaron Courville, and Yoshua Bengio. Generative adversarial nets. In *Advances in neural information processing systems*, pages 2672–2680, 2014.
14. Yann A. LeCun, Yoshua Bengio, and Geoffrey E. Hinton. Deep learning. *Nature*, 2015.
15. Geoffrey Hinton, Li Deng, Dong Yu, George E Dahl, Abdel-rahman Mohamed, Navdeep Jaitly, Andrew Senior, Vincent Vanhoucke, Patrick Nguyen, Tara N Sainath, and Brian Kingsbury. Deep Neural Networks for Acoustic Modeling in Speech Recognition. *Ieee Signal Processing Magazine*, 2012.
16. Kaiming He, Xiangyu Zhang, Shaoqing Ren, and Jian Sun. Deep Residual Learning for Image Recognition. In *2016 IEEE Conference on Computer Vision and Pattern Recognition (CVPR)*, 2016.
17. John R. Koza, Forrest H. Bennett, David Andre, and Martin A. Keane. Automated Design of Both the Topology and Sizing of Analog Electrical Circuits Using Genetic Programming. In *Artificial Intelligence in Design '96*. 1996.
18. Li-Ming Fu. *Neural networks in computer intelligence*. Tata McGraw-Hill Education, 2003.
19. M. I. Jordan and T. M. Mitchell. Machine learning: Trends, perspectives, and prospects, 2015.
20. Lucian Busoniu, Robert Babuska, Bart De Schutter, and Damien Ernst. *Reinforcement learning and dynamic programming using function approximators*. CRC press, 2010.
21. Thomas Jaksch, Ronald Ortner, and Peter Auer. Near-optimal regret bounds for reinforcement learning. *Journal of Machine Learning Research*, 11(Apr):1563–1600, 2010.
22. Dimitri P Bertsekas and John N Tsitsiklis. Neuro-dynamic programming: an overview. In *Proceedings of the 34th IEEE Conference on Decision and Control*, volume 1, pages 560–564. IEEE Publ. Piscataway, NJ, 1995.
23. Stuart J. Russell and Peter Norvig. *Artificial Intelligence: A Modern Approach*. 3rd edition, 2015.
24. D. L. Poole, Alan Mackworth, and R. G. Goebel. Computational Intelligence and Knowledge. *Computational Intelligence: A Logical Approach*, (Ci):1–22, 1998.
25. Ian H. Witten, Eibe Frank, and Mark a. Hall. *Data Mining: Practical Machine Learning Tools and Techniques, Third Edition*, volume 54. 2011.
26. Pete Chapman, Julian Clinton, Randy Kerber, Thomas Khabaza, Thomas Reinartz, Colin Shearer, and Rudiger Wirth. Crisp-Dm 1.0. *CRISP-DM Consortium*, page 76, 2000.
27. Lucas Baier, Niklas Kühl, and Gerhard Satzger. How to Cope with Change? Preserving Validity of Predictive Services over Time. In *Hawaii International Conference on System Sciences (HICSS-52)*, Grand Wailea, Maui, Hawaii, USA, 2019.
28. Franziska Oroszi and Johannes Ruhland. An early warning system for hospital acquired. In *18th European Conference on Information Systems (ECIS)*, 2010.
29. Xiaofeng Yang, Jian Su, and Chew Lim Tan. A twin-candidate model for learning-based anaphora resolution. *Computational Linguistics*, 34(3):327–356, 2008.

30. Zhedong Zheng, Liang Zheng, and Yi Yang. Pedestrian alignment network for large-scale person re-identification. *arXiv preprint arXiv:1707.00408*, 2017.
31. Robin Hirt, Niklas Kühl, and Gerhard Satzger. An end-to-end process model for supervised machine learning classification: from problem to deployment in information systems. In *Proceedings of the DESRIST 2017 Research-in-Progress*. Karlsruher Institut für Technologie (KIT), 2017.
32. J Rowley and B Keegan. Looking back, going forward: the role and nature of systematic literature reviews in digital marketing: a meta-analysis. 2017.
33. Peter JH Hulshof, Nikky Kortbeek, Richard J Boucherie, Erwin W Hans, and Piet JM Bakker. Taxonomic classification of planning decisions in health care: a structured review of the state of the art in or/ms. *Health systems*, 1(2):129–175, 2012.
34. Ameen Abu-Hanna and Nicolette de Keizer. Integrating classification trees with local logistic regression in intensive care prognosis. *Artificial Intelligence in Medicine*, 29(1–2):5–23, 2003.
35. C William Hanson and Bryan E Marshall. Artificial intelligence applications in the intensive care unit. *Critical care medicine*, 29(2):427–435, 2001.
36. Aditya Khosla, Yu Cao, Cliff Chiung-Yu Lin, Hsu-Kuang Chiu, Junling Hu, and Honglak Lee. An integrated machine learning approach to stroke prediction. In *Proceedings of the 16th ACM SIGKDD international conference on Knowledge discovery and data mining*, pages 183–192. ACM, 2010.
37. Klaus Donsa, Stephan Spat, Peter Beck, Thomas R Pieber, and Andreas Holzinger. Towards personalization of diabetes therapy using computerized decision support and machine learning: some open problems and challenges. In *Smart Health*, pages 237–260. Springer, 2015.
38. Ahmad Hammoudeh, Ghazi Al-Naymat, Ibrahim Ghannam, and Nadim Obied. Predicting hospital readmission among diabetics using deep learning. *Procedia Computer Science*, 141:484–489, 2018.
39. Mu Zhu, Zhanyang Zhang, John P Hirdes, and Paul Stolee. Using machine learning algorithms to guide rehabilitation planning for home care clients. *BMC medical informatics and decision making*, 7(1):41, 2007.
40. Jinsung Yoon, Camelia Davtyan, and Mihaela van der Schaar. Discovery and clinical decision support for personalized healthcare. *IEEE journal of biomedical and health informatics*, 21(4):1133–1145, 2017.
41. Danica Mitch M Pacis, Edwin DC Subido Jr, and Nilo T Bugtai. Trends in telemedicine utilizing artificial intelligence. In *AIP Conference Proceedings*, volume 1933, page 040009. AIP Publishing, 2018.
42. Stephan Kudyba, G Brent Hamar, and William M Gandy. Utilising neural network applications to enhance efficiency in the healthcare industry: predicting populations of future chronic illness. *International Journal of Business Intelligence and Data Mining*, 1(4):371–383, 2006.
43. Paras Lakhani, Adam B Prater, R Kent Hutson, Kathy P Andriole, Keith J Dreyer, Jose Morey, Luciano M Prevedello, Toshi J Clark, J Raymond Geis, Jason N Itri, et al. Machine learning in radiology: applications beyond image interpretation. *Journal of the American College of Radiology*, 15(2):350–359, 2018.
44. Gilmer Valdes, Charles B Simone II, Josephine Chen, Alexander Lin, Sue S Yom, Adam J Pattison, Colin M Carpenter, and Timothy D Solberg. Clinical decision support of radiotherapy treatment planning: A data-driven machine learning strategy for patient-specific dosimetric decision making. *Radiotherapy and Oncology*, 125(3):392–397, 2017.
45. Fei Jiang, Yong Jiang, Hui Zhi, Yi Dong, Hao Li, Sufeng Ma, Yilong Wang, Qiang Dong, Haipeng Shen, and Yongjun Wang. Artificial intelligence in healthcare: past, present and future. *Stroke and vascular neurology*, 2(4):230–243, 2017.
46. Constantine D Spyropoulos. Ai planning and scheduling in the medical hospital environment, 2000.

47. Illhoi Yoo, Patricia Alafaireet, Miroslav Marinov, Keila Pena-Hernandez, Rajitha Gopidi, Jia-Fu Chang, and Lei Hua. Data mining in healthcare and biomedicine: a survey of the literature. *Journal of medical systems*, 36(4):2431–2448, 2012.
48. Parvez Ahmad, Saqib Qamar, and Syed Qasim Afser Rizvi. Techniques of data mining in healthcare: a review. *International Journal of Computer Applications*, 120(15), 2015.
49. Manoj Durairaj and Veera Ranjani. Data mining applications in healthcare sector: a study. *International journal of scientific & technology research*, 2(10):29–35, 2013.
50. Hian Chye Koh, Gerald Tan, et al. Data mining applications in healthcare. *Journal of healthcare information management*, 19(2):65, 2011.
51. Tanvi Anand, Rekha Pal, and Sanjay Kumar Dubey. Data mining in healthcare informatics: Techniques and applications. In *2016 3rd International Conference on Computing for Sustainable Global Development (INDIACom)*, pages 4023–4029. IEEE, 2016.
52. Divya Tomar and Sonali Agarwal. A survey on data mining approaches for healthcare. *International Journal of Bio-Science and Bio-Technology*, 5(5):241–266, 2013.
53. MM Malik, S Abdallah, and M Alaâ raj. Data mining and predictive analytics applications for the delivery of healthcare services: a systematic literature review. *Annals of Operations Research*, 270(1–2):287–312, 2018.
54. Venketesh Palanisamy and Ramkumar Thirunavukarasu. Implications of big data analytics in developing healthcare frameworks–a review. *Journal of King Saud University-Computer and Information Sciences*, 2017.
55. Maria Antonina Mach and M Salem Abdel-Badeeh. Intelligent techniques for business intelligence in healthcare. In *2010 10th International Conference on Intelligent Systems Design and Applications*, pages 545–550. IEEE, 2010.
56. Nishita Mehta, Anil Pandit, and Sharvari Shukla. Transforming healthcare with big data analytics and artificial intelligence: A systematic mapping study. *Journal of biomedical informatics*, page 103311, 2019.
57. Nida Shahid, Tim Rappon, and Whitney Berta. Applications of artificial neural networks in health care organizational decision-making: A scoping review. *PloS one*, 14(2):e0212356, 2019.
58. Jean-Emmanuel Bibault, Philippe Giraud, and Anita Burgun. Big data and machine learning in radiation oncology: state of the art and future prospects. *Cancer letters*, 382(1):110–117, 2016.
59. Jaleh Shoshtarian Malak, Hojjat Zeraati, Fatemeh Sadat Nayeri, Reza Safdari, and Azimeh Danesh Shahraki. Neonatal intensive care decision support systems using artificial intelligence techniques: a systematic review. *Artificial Intelligence Review*, 52(4):2685–2704, 2019.
60. Albert Y Chen, Tsung-Yu Lu, Matthew Huei-Ming Ma, and Wei-Zen Sun. Demand forecast using data analytics for the preallocation of ambulances. *IEEE journal of biomedical and health informatics*, 20(4):1178–1187, 2016.
61. Hisashi Kurasawa, Katsuyoshi Hayashi, Akinori Fujino, Koichi Takasugi, Tsuneyuki Haga, Kayo Waki, Takashi Noguchi, and Kazuhiko Ohe. Machine-learning-based prediction of a missed scheduled clinical appointment by patients with diabetes. *Journal of diabetes science and technology*, 10(3):730–736, 2016.
62. Nathan Benjamin Menke, Nicholas Caputo, Robert Fraser, Jordana Haber, Christopher Shields, and Marie Nam Menke. A retrospective analysis of the utility of an artificial neural network to predicted volume. *The American journal of emergency medicine*, 32(6):614–617, 2014.
63. Mohamed Afilal, Farouk Yalaoui, Frédéric Dugardin, Lionel Amodeo, David Laplanche, and Philippe Blua. Emergency department flow: A new practical patients classification and forecasting daily attendance. *IFAC-PapersOnLine*, 49(12):721–726, 2016.
64. Mai Xu, Tse Chiu Wong, and Kwai-Sang Chin. Modeling daily patient arrivals at emergency department and quantifying the relative importance of contributing variables using artificial neural network. *Decision Support Systems*, 54(3):1488–1498, 2013.

65. Silvia Acid, Luis M de Campos, Juan M Fernández-Luna, Susana Rodrıguez, José Marıa Rodrıguez, and José Luis Salcedo. A comparison of learning algorithms for Bayesian networks: a case study based on data from an emergency medical service. *Artificial intelligence in medicine*, 30(3):215–232, 2004.
66. Byron Graham, Raymond Bond, Michael Quinn, and Maurice Mulvenna. Using data mining to predict hospital admissions from the emergency department. *IEEE Access*, 6:10458–10469, 2018.
67. Jonas Krämer, Jonas Schreyögg, and Reinhard Busse. Classification of hospital admissions into emergency and elective care: a machine learning approach. *Health care management science*, 22(1):85–105, 2019.
68. Seung-Yup Lee, Ratna Babu Chinnam, Evrim Dalkiran, Seth Krupp, and Michael Nauss. Prediction of emergency department patient disposition decision for proactive resource allocation for admission. *Health care management science*, pages 1–21, 2019.
69. Sofia Benbelkacem, Farid Kadri, Baghdad Atmani, and Sondès Chaabane. Machine learning for emergency department management. *International Journal of Information Systems in the Service Sector (IJISSS)*, 11(3):19–36, 2019.
70. Xiongcai Cai, Oscar Perez-Concha, Enrico Coiera, Fernando Martin-Sanchez, Richard Day, David Roffe, and Blanca Gallego. Real-time prediction of mortality, readmission, and length of stay using electronic health record data. *Journal of the American Medical Informatics Association*, 23(3):553–561, 2015.
71. Li Luo, Jialing Li, Chuang Liu, and Wenwu Shen. Using machine-learning methods to support health-care professionals in making admission decisions. *The International journal of health planning and management*, 2019.
72. Lior Turgeman and Jerrold H May. A mixed-ensemble model for hospital readmission. *Artificial intelligence in medicine*, 72:72–82, 2016.
73. Bichen Zheng, Jinghe Zhang, Sang Won Yoon, Sarah S Lam, Mohammad Khasawneh, and Srikanth Poranki. Predictive modeling of hospital readmissions using metaheuristics and data mining. *Expert Systems with Applications*, 42(20):7110–7120, 2015.
74. Eva K Lee, Fan Yuan, Daniel A Hirsh, Michael D Mallory, and Harold K Simon. A clinical decision tool for predicting patient care characteristics: patients returning within 72 hours in the emergency department. In *AMIA Annual Symposium Proceedings*, volume 2012, page 495. American Medical Informatics Association, 2012.
75. Joseph Futoma, Jonathan Morris, and Joseph Lucas. A comparison of models for predicting early hospital readmissions. *Journal of Biomedical Informatics*, 56:229–238, 2015.
76. Santu Rana, Sunil Gupta, Dinh Phung, and Svetha Venkatesh. A predictive framework for modeling healthcare data with evolving clinical interventions. *Statistical Analysis and Data Mining: The ASA Data Science Journal*, 8(3):162–182, 2015.
77. Isabella Eigner, Freimut Bodendorf, and Nilmini Wickramasinghe. A theoretical framework for research on readmission risk prediction. In *32ND Bled eConference Humanizing Technology for a Sustainable Society*, pages 387–410, 2019.
78. Nilmini Wickramasinghe, Day Manuet Delgano, and Steven Mcconchie. Real-time prediction of the risk of hospital readmissions. In *32ND Bled eConference Humanizing Technology for a Sustainable Society*, pages 85–102, 2019.
79. Christopher J McWilliams, Daniel J Lawson, Raul Santos-Rodriguez, Iain D Gilchrist, Alan Champneys, Timothy H Gould, Mathew JC Thomas, and Christopher P Bourdeaux. Towards a decision support tool for intensive care discharge: machine learning algorithm development using electronic healthcare data from MIMIC-III and Bristol, UK. *BMJ open*, 9(3):e025925, 2019.
80. Nicolas Padoy. Machine and deep learning for workflow recognition during surgery. *Minimally Invasive Therapy & Allied Technologies*, 28(2):82–90, 2019.
81. Anastasia A Funkner, Aleksey N Yakovlev, and Sergey V Kovalchuk. Data-driven modeling of clinical pathways using electronic health records. *Procedia computer science*, 121:835–842, 2017.

82. Noura Al Nuaimi. Data mining approaches for predicting demand for healthcare services in Abu Dhabi. In *2014 10th International Conference on Innovations in Information Technology (IIT)*, pages 42–47. IEEE, 2014.
83. Kazim Topuz, Hasmet Uner, Asil Oztekin, and Mehmet Bayram Yildirim. Predicting pediatric clinic no-shows: a decision analytic framework using elastic net and Bayesian belief network. *Annals of Operations Research*, 263(1–2):479–499, 2018.
84. Amy Nelson, Daniel Herron, Geraint Rees, and Parashkev Nachev. Predicting scheduled hospital attendance with artificial intelligence. *Npj Digital Medicine*, 2(1):26, 2019.
85. Rachel M Goffman, Shannon L Harris, Jerrold H May, Aleksandra S Milicevic, Robert J Monte, Larissa Myaskovsky, Keri L Rodriguez, Youxu C Tjader, and Dominic L Vargas. Modeling patient no-show history and predicting future outpatient appointment behavior in the Veterans Health Administration. *Military medicine*, 182(5–6):e1708–e1714, 2017.
86. Shannon L Harris, Jerrold H May, and Luis G Vargas. Predictive analytics model for healthcare planning and scheduling. *European Journal of Operational Research*, 253(1):121–131, 2016.
87. Adel Alaeddini, Kai Yang, Pamela Reeves, and Chandan K Reddy. A hybrid prediction model for no-shows and cancellations of outpatient appointments. *IIE Transactions on healthcare systems engineering*, 5(1):14–32, 2015.
88. Lee Goldman, Ralph Freidin, E Francis Cook, John Eigner, and Pamela Grich. A multivariate approach to the prediction of no-show behavior in a primary care center. *Archives of Internal Medicine*, 142(3):563–567, 1982.
89. Isabella Eigner, Freimut Bodendorf, and Nilmini Wickramasinghe. Predicting high-cost patients by machine learning: A case study in an Australian private hospital group. In *Proceedings of 11th International Conference*, volume 60, pages 94–103, 2019.
90. Zhengyi Zhou. Predicting ambulance demand: Challenges and methods. *arXiv preprint arXiv:1606.05363*, 2016.
91. Albert Y Chen and Tsung-Yu Lu. A GIS-based demand forecast using machine learning for emergency medical services. *Computing in Civil and Building Engineering (2014)*, pages 1634–1641, 2014.
92. Melanie Villani, Arul Earnest, Natalie Nanayakkara, Karen Smith, Barbora De Courten, and Sophia Zoungas. Time series modelling to forecast prehospital EMS demand for diabetic emergencies. *BMC health services research*, 17(1):332, 2017.
93. Catherine Curtis, Chang Liu, Thomas J Bollerman, and Oleg S Pianykh. Machine learning for predicting patient wait times and appointment delays. *Journal of the American College of Radiology*, 15(9):1310–1316, 2018.
94. Ines Verena Arnolds and Daniel Gartner. Improving hospital layout planning through clinical pathway mining. *Annals of Operations Research*, 263(1–2):453–477, 2018.
95. Daniel Gartner, Rainer Kolisch, Daniel B Neill, and Rema Padman. Machine learning approaches for early DRG classification and resource allocation. *INFORMS Journal on Computing*, 27(4):718–734, 2015.
96. J Alapont, A Bella-Sanjuán, C Ferri, J Hernández-Orallo, JD Llopis-Llopis, MJ Ramírez-Quintana, et al. Specialised tools for automating data mining for hospital management. In *Proc. First East European Conference on Health Care Modelling and Computation*, pages 7–19, 2005.
97. Kottalanka Srikanth and D Arivazhagan. An efficient patient inflow prediction model for hospital resource management. *ICTACT Journal on Soft Computing*, 7(4), 2017.
98. Emel Aktaş, Füsun Ülengin, and Şule Önsel Şahin. A decision support system to improve the efficiency of resource allocation in healthcare management. *Socio-Economic Planning Sciences*, 41(2):130–146, 2007.
99. Marek Laskowski. A prototype agent based model and machine learning hybrid system for healthcare decision support. *International Journal of E-Health and Medical Communications (IJEHMC)*, 2(4):67–90, 2011.

100. R Ceglowski, Leonid Churilov, and J Wasserthiel. Combining data mining and discrete event simulation for a value-added view of a hospital emergency department. *Journal of the Operational Research Society*, 58(2):246–254, 2007.
101. Reid F Thompson, Gilmer Valdes, Clifton D Fuller, Colin M Carpenter, Olivier Morin, Sanjay Aneja, William D Lindsay, Hugo JWL Aerts, Barbara Agrimson, Curtiland Deville, et al. Artificial intelligence in radiation oncology: a specialty-wide disruptive transformation? *Radiotherapy and Oncology*, 2018.
102. Vahid Lotfi and Edgar Torres. Improving an outpatient clinic utilization using decision analysis-based patient scheduling. *Socio-Economic Planning Sciences*, 48(2):115–126, 2014.
103. Sharan Srinivas and A Ravi Ravindran. Optimizing outpatient appointment system using machine learning algorithms and scheduling rules: A prescriptive analytics framework. *Expert Systems with Applications*, 102:245–261, 2018.
104. Paul Harper. Combining data mining tools with health care models for improved understanding of health processes and resource utilisation. *Clinical and investigative medicine*, 28(6):338, 2005.
105. George Grekousis and Ye Liu. Where will the next emergency event occur? predicting ambulance demand in emergency medical services using artificial intelligence. *Computers, Environment and Urban Systems*, 76:110–122, 2019.
106. Adam T Zagorecki, David EA Johnson, and Jozef Ristvej. Data mining and machine learning in the context of disaster and crisis management. *International Journal of Emergency Management*, 9(4):351–365, 2013.
107. Iulia Lefter, Leon JM Rothkrantz, David A Van Leeuwen, and Pascal Wiggers. Automatic stress detection in emergency(telephone) calls. *International Journal of Intelligent Defence Support Systems*, 4(2):148–168, 2011.
108. Benjamin Potter. *Constructing Efficient Production Networks: A Machine Learning Approach*. PhD thesis, Department of Mechanical and Industrial Engineering, University of Toronto, 2018.
109. Jonathan Strahl. Patient appointment scheduling system: with supervised learning prediction, 2015.
110. Josefine Letzner. Analysis of emergency medical transport datasets using machine learning, 2017.
111. Barbara Kitchenham, O Pearl Brereton, David Budgen, Mark Turner, John Bailey, and Stephen Linkman. Systematic literature reviews in software engineering–a systematic literature review. *Information and software technology*, 51(1):7–15, 2009.
112. Philip M Podsakoff, Scott B MacKenzie, Jeong-Yeon Lee, and Nathan P Podsakoff. Common method biases in behavioral research: A critical review of the literature and recommended remedies. *Journal of applied psychology*, 88(5):879, 2003.

Chapter 2
AI/OR Synergies of Process Mining with Optimal Planning of Patient Pathways for Effective Hospital-Wide Decision Support

Taïeb Mellouli and Thomas Stoeck

Abstract Within a hospital, most of the relevant decision problems focus on patient pathways. Earlier studies show that OR techniques are capable of planning these heterogeneous pathways in multiple ward bed management. This chapter focuses on the synergies of these OR techniques with AI process mining. The first type of AI/OR synergy is to adapt process mining to learn the structures of patient pathways for different diagnoses and forward them to optimization. A case study with yearly data of a university hospital shows many business benefits and better planning results quality for shared resources. Further, the second type of AI/OR synergy is detected after a profound study of problem classes within our two-dimensional classification scheme considering both complexity dimensions related to patient pathways and hospital resources. Because of the many hard-to-model interdependencies, model-reality gaps occur. For the second type of AI/OR synergy, process mining is interpreted in a dual way where discrepancies of operations against planned pathway structure are detected. For this sake, a discrepancy mining approach is proposed and embedded with optimization into a plan-and-refine framework.

2.1 Motivation and Research Outline

Facing unprecedented challenges and economic pressure, hospitals are forced to make their processes more effective and to use their resources more efficiently. A serious burden of effective decision support in hospitals for all actors and decision makers, like physicians, nurses, and hospital managers, is the simultaneous twofold complexity inherent in their professional planning and organizational decisions. By clarifying the two interrelated dimensions of complexity, their interplay, and the many interdependencies to be considered for achieving better expert decisions, we

T. Mellouli (✉) · T. Stoeck
Department of Business Information Systems and Operations Research, Martin Luther University Halle-Wittenberg, Halle, Germany
e-mail: mellouli@wiwi.uni-halle.de; thomas.stoeck@wiwi.uni-halle.de

M. Masmoudi et al. (eds.), *Artificial Intelligence and Data Mining in Healthcare*,
https://doi.org/10.1007/978-3-030-45240-7_2

show in this paper many facets of synergy between descriptive/predictive mining of complex patient pathways and process discrepancies within hospitals from an AI point of view and prescriptive planning of these complex entities considering complex resource requirements in hospitals from an OR point of view.

2.1.1 AI/OR Synergies meet Hospital Decision Task Complexities

The first type/dimension of complexity is related to the complex composition of clinical pathways, which patients should "follow" through the hospital over different wards depending on the locations of their associated treatments. A clinical pathway associated to a certain pathology/diagnosis contains detailed information about both clinically relevant treatments and organizationally relevant allocation of beds in hospital's wards. Besides physicians' knowledge about the treatment package for a patient with certain pathology, AI methods for descriptive/predictive process mining of historical hospital data may help them to decide about details of the clinical pathways to be adopted in operations. As prescriptive planning of groups of patients in one or several ward(s) cannot be realistic without accurate knowledge of the patient pathways, a first type of synergy between pathway mining and prescriptive planning arises: By extracting relevant pathway structure and pathway constraints from a process mining procedure, we are able to utilize them as input data and for the formulation of constraints within the mathematical models for scheduling patient pathways in wards. Thus, the first type of synergy is to utilize process mining results of pathways in prescriptive planning methods. We illustrate the benefits of this first type of synergy by considering case studies based on yearly hospital data and previous works published in an AI-oriented and an OR-oriented paper of our research group.

The second type/dimension of complexity is related to the complex environment of resources and their organizational management demarcated by the existence of several wards within hospitals with shared resources and with several other types of hidden interdependencies. First, we distinguish and discuss in some details six types of these interdependencies that have to be taken into account for accurate decision-making and decision support in hospitals. Second, we combine two dimensions of grads of consideration (or modeling granularities) of the first and second type of complexity into a two-dimensional problem scheme in order to demarcate different (aggregation levels of) decision-making and decision support tasks of hospital expert and management actors. Based on this two-dimensional scheme for decision support in hospitals, we are able to clearly differentiate various AI-/mining-oriented tasks and prescriptive planning OR-oriented tasks within hospital decisions, to detect AI/AI process/data mining synergies, and AI/OR synergies similar to the abovementioned first type. Moreover, we illustrate the substantial gaps between modeling and reality when considering ways for effective hospital-wide decision support. These model–reality gaps are generally existent, because not all types of interdependency could be considered in prescriptive planning models and methods. Some types of interdependencies remain outside modeling consideration in spite of integrating heterogeneous pathway flows in different wards with shared capacities within (hierarchical) mathematical optimization and simulation models in our research.

In order to minimize these model–reality gaps, we introduce a second kind of synergy of predictive mining and prescriptive planning of complex pathways for effective hospital-wide decision support in hospitals. To operationalize this second type of AI/OR synergy, we propose a new kind of mining procedures of process discrepancies in hospitals caused by the model–reality gaps, in order to learn strain situations in hospitals leading to disagreement between result plans of prescriptive methods and operational reality. Ideally, the results of this mining approach would resemble expert rules describing bottleneck situations in terms of general modeling entities, so that these learned rules could be integrated into prescriptive models. As this method is not likely to exhaustively handle all types of interdependencies, individual instantiated bottleneck and local strain situations recognized within plans of optimization results can be utilized to iteratively add restrictions to the prescriptive model (cuts) in order to avoid these situations in next optimization rounds.

This subsection summarizes the argumentation path and the main contents of the chapter. The last three paragraphs roughly correspond to synopses of the three main sections of this chapter. Note that most of the new contributions are presented in Sects. 2.3 and 2.4 after discussing case studies in Sect. 2.2 in a business-oriented manner showing economic benefits in practice.

2.1.2 Pathway Centered Decision Support Toward AI/OR Synergy

After introducing DRG (diagnosis-related groups) payment systems for clinical services in many countries, notably the German DRG (G-DRG) in 2003 in Germany, hospitals are no longer payed on the basis of patient length of stay and actual treatment services offered, but rather on a diagnosis basis with associated standard treatment packages and standard fixed values of patient length of stay independent of hospital choice. Under these circumstances, economic pressure on hospitals is rising and they are facing unprecedented challenges in management, accounting, control, and decision-making. Hospitals are, therefore, forced to make their processes more effective and to use their resources more efficiently.

In enterprises with complex structures both in terms of their products and services as well as their organizational resources, decision-making and decision support tasks are interrelated and hide many difficulties in terms of modeling and solution approaches. Complex structures arise in hospitals for the services to be offered to patients in form of clinical pathways. According to [6], the at most leverage effect in hospital performance is promised by the establishment of patient-centered clinical pathways. Referring to [7], a patient (or clinical) pathway is a multidisciplinary plan determining which treatment stations/wards a patient pass though from admission until discharge. In fact, clinical pathways contain detailed information about both clinically relevant treatments and organizationally relevant allocation of beds in hospital's wards where treatments and other services are offered.

Patient pathways help hospitals to comply with DRG regulations, to effectively read just their processes, and to efficiently seize their resource usage. The level of details offered by patient pathways enables hospitals to manage their processes effectively and control their costs accurately. Relevant information about treatments within these clinical pathways are stored on a regular basis in order to ensure compliance to government directives with respect to DRG and treatment standards. Information on pathways also helps insurances to compare treatment practice of hospitals depending on patient diagnosis and therefore obtain accounting transparency. On this basis, a fair comparison of clinics and hospitals in terms of performance for special diagnoses is possible.

In light of the many management and controlling facets of the information about clinical pathways, a new nationwide standard of these data is introduced in Germany (known as standard §21 data) and is continuously being refined together with adjusting treatment standards and treatment costs for diagnosis groups as directives for hospitals (see [8, 9]). Our research group at the department of business information systems and operations research at the Martin Luther University Halle-Wittenberg (headed by the first author) recognized the potentials of these standard management and accounting data (§21 data) in order to deliver effective decision support for hospitals, ideally in generic way for German hospitals. Five years ago [2], a simulation model for hospital-wide clinical pathway flows (modeling paths of patient groups across wards) is realized in order to evaluate flexible bed allocation scenarios on the basis of yearly data of a university hospital. The simulation model, initially realized in ARENA, is reimplemented in a generic way [3] using the object-oriented simulation environment SIMIO in order to evaluate flexible ward clusters in hospital occupancy management.

In order to deliver accurate information on pathways for detailed prescriptive planning and scenario simulation, a clinical pathway mining approach is presented [4] in order to enable scheduling of hospital relocations and treatment services. It is argued that, for an effective use of mining results in predictive planning methods, information on pathway structure as well as pathway restrictions should be mined in order to be integrated in scenario simulations of pathway flows through the hospital [3] and in optimization methods for scheduling pathways [5]. In [5], it is shown that the information on pathway structure and pathway restrictions and rules gained by process mining can be effectively used in prescriptive planning of clinical pathways using mathematical optimization giving rise to a data-driven hierarchical mixed-integer programming approach. A real-world case of German university hospital department shows the benefits on the solution quality of prescriptive planning.

2.1.3 Research on AI/OR Synergy and Chapter Outline

In this chapter, we first review in the next section these research works and case studies in order to introduce and practically introduce a first type of AI/OR synergy within a framework for hospital decision support. With this type of AI/OR synergy, utilizing the pathway structures gained by process mining increases the accuracy of prescriptive planning of bed-and-ward allocation for patients.

The second part of the chapter includes two sections: the third section is devoted to detection types and instances of AI/OR synergies within hospital decision support and the last fourth section to clarify the new detected second type of AI/OR synergy and to propose ways to operationalize this synergy, which is demarcated by mining of process discrepancies and its interplay with prescriptive planning toward effective hospital-wide decision support. The third and fourth sections are based on recent works and presentations of the authors. The first fundamental work [1] proposes and discusses a two-dimensional categorization scheme for simulation-/optimization-based decision support in hospitals applied to overall bed management in interdependent wards under flexibility. The first research ideas exposed in this chapter were presented at the "Digital Agenda Workshop" on Artificial Intelligence organized by the ministry of economics, sciences, and digitalization of the Land Saxony-Anhalt in cooperation with "Leopoldina" - Germany's National Academy of Sciences.[1] In this workshop, our department participates with 5 presentations (out of a total of 14), distributed into three sessions: three contributions within the "intelligent city" session, and one to each of the sessions "intelligent industry" and "intelligent society." Two of the three contributions within the "intelligent city" session are given by the authors and concerned "synergy between predictive AI and prescriptive planning methods for effective decision support in hospitals" and "AI-based mining of clinical pathways and process discrepancies for an anticipating self-learning planning in hospitals."

The second part of the paper opens new directions of research on AI/OR synergy by discussing new ideas of these two AI Workshop presentations with reference to fundamental work on the two-dimensional scheme in [1]. As sketched in the Abstract/Summary, it begins in Sect. 2.3 with a detailed discussion of interdependencies in hospitals affecting decision-making and decision support tasks. Then two orthogonal dimensions of complexity are differentiated, one of them related to the composition of pathways and the other to the complex environment with respect to interdependent wards and shared resources and organizations in hospitals. Based on this, a two-dimensional scheme is proposed to categorize various decision-making and decision support tasks in hospitals. Two types of AI/OR synergy are then detected within the two-dimensional scheme: The first straightforward AI/OR synergy is the connection between a single pathway consideration of process

[1]Digitale Agenda Workshop künstliche Intelligenz. Kooperationsveranstaltung des Ministeriums für Wirtschaft, Wissenschaft und Digitalisierung mit der Leopoldina—Nationale Akademie der Wissenschaften, Halle, August 29, 2019.

mining and a multiple pathway flow needed for resource and bed allocation for patient groups in one or several wards (homogenous or heterogenous pathways) in prescriptive planning methods. The second new type of AI/OR synergy is presented and discussed in more detail in Sect. 2.4. The idea is to propose a second kind of AI/mining procedure devoted to process discrepancies and to show its interplay with prescriptive planning toward effective hospital-wide decision support. Referring to the aggregation level of pathway composition in the two-dimensional scheme, both types of AI/OR synergy could be viewed at different level of details: aggregated resource and bed allocation level as well as treatment package and/or treatment sequence level.

2.2 First Type of AI/OR Synergy: Process Mining of Pathways for Accurate Prescriptive Planning of Ward-and-Bed Allocation

In this section, we illustrate by decision-making and decision support scenarios in hospitals, how descriptive/predictive analytics methods both for simple and complex structures have benefits toward a more accurate prescriptive planning of patient pathways and ward-and-bed allocations in hospitals. Realistic case studies discussed in previous works of our department [4, 5] show the impact of the resulting AI/OR synergy (of first type). This business-oriented exposition of this section helps the reader to follow the more fundamental two-dimensional categorization in the next section and the more abstract and generic discussion of types of AI/OR synergies.

2.2.1 Synergy between Predictive and Prescriptive Analytics: Cases of Simple vs. Complex Structures

While descriptive and predictive analytics help the manager and decision maker to be informed on what is happened and to analyze what will happen from an outside observer point of view, prescriptive methods support decision maker to shape the future by optimizing plans for operations from an inside decision maker point of view. Major prescriptive planning tasks in hospitals considered in our research are linked to ward-and-bed allocation for inpatients (patients with planned hospital stays) and to the evaluation of scenarios of flexible resource usage for inpatients from both flows of elective (planned-stay) patients and of emergency patients (forwarded to hospital wards). Prescriptive OR methods support decisions in both cases, in the first kind of tasks, by mathematical optimization approaches (e.g., [5]) and for the second kind of tasks, by simulation models helping to evaluate different organizational scenarios (see, e.g., [4]). For both prescriptive analytics approaches, we need accurate input data, parameter, distributions, KPI,

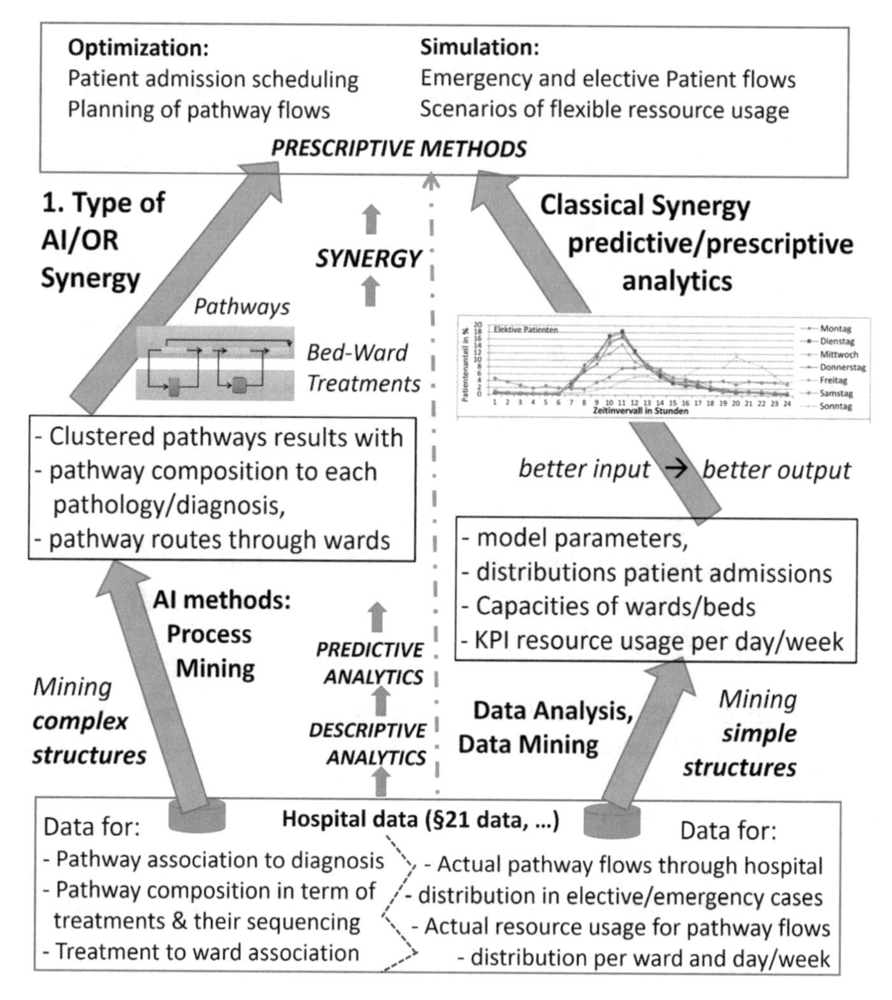

Fig. 2.1 Synergies between descriptive/predictive and prescriptive analytics for hospital decision support: classical synergy and 1. Type of AI/OR synergy

etc., in order to get accurate planning and scenario evaluation results. The better the input quality, the better the output quality. This is depicted in Fig. 2.1 (right side arrows) and denominated as the classical synergy between descriptive/predictive and prescriptive analytics.

For this classical synergy, simple structures in the form of model parameter, distributions of patient admissions, capacities of ward bed, and KPI for resource usage per day/week are needed as input for optimization and simulation models of prescriptive methods. Data analysis and simple data mining tools help gathering these simple structures first in a descriptive analytics framework. Taking into account trends and factors influencing the future environment (also depending on scenarios to be evaluated), predictive results for these simple structures are gained.

In prescriptive methods, where complex structures of objects/entities/tasks like clinical pathways are planned, the composition of these structures and other parameters concerning these structures should be known in advance as input. It is rarely the case that experts know all needed details about these complex structures. A way to gather the complex structures is to apply predictive AI methods such as pattern recognition, machine learning, association rule mining, or process mining. Thus, another form of synergy between descriptive/predictive and prescriptive analytics is needed, which is depicted in Fig. 2.1 (left side arrows) and denominated as "1. Type of AI/OR synergy." For the hospital case study, a special process mining approach for clinical pathways that harmonizes with the needs of the prescriptive optimization models to plan pathway flows in wards is proposed. This special approach for process mining is illustrated in the next subsection, and the benefits of the AI/OR synergy are shown by considering case studies with data of a university hospital.

2.2.2 *First Type of AI/OR Synergy and Its Benefits for Effective Hospital Decision Support: Case Study of a University Hospital*

A special clinical pathway mining approach is presented in [4]. In order to enable accurate prescriptive scheduling of hospital relocations and treatment services, it is argued that information on pathway structure and pathway restrictions are needed in order to be integrated into the prescriptive optimization model (see Fig. 2.2). To each clustered pathology-dependent pathway class, the pathway structure indicates two detailed levels (the left side of Fig. 2.2). The first level indicates the succession of wards that can be taken by a patient following the corresponding clustered and generalized pathway, where possible alternative wards can be chosen. Anchored with the ward-day indications of the first level, the second level specifies which treatments are to be offered depending on the treatment package associated with the pathway. Also here, it is possible to choose between alternative treatments within the pathway—a good opportunity for optimization depending on availability of resources. As the pathways gained by process mining are considered in clusters, the variability of the length of stay at the ward-day level and treatment occurrence on the treatments level are indicated by the variables min, max, average, and median. The pathway constraints on the right side of Fig. 2.2 include the corresponding tables/information to these variabilities together with order constraints between ward-day relocations and between treatments as well as anchored constraints between the two kinds of these time-lasting "activities." For these order constraints, the transition probability is indicated together with lag-bandwidth (min lag, max lag) for the allowed start of successor activity/step with respect to the start of considered step.

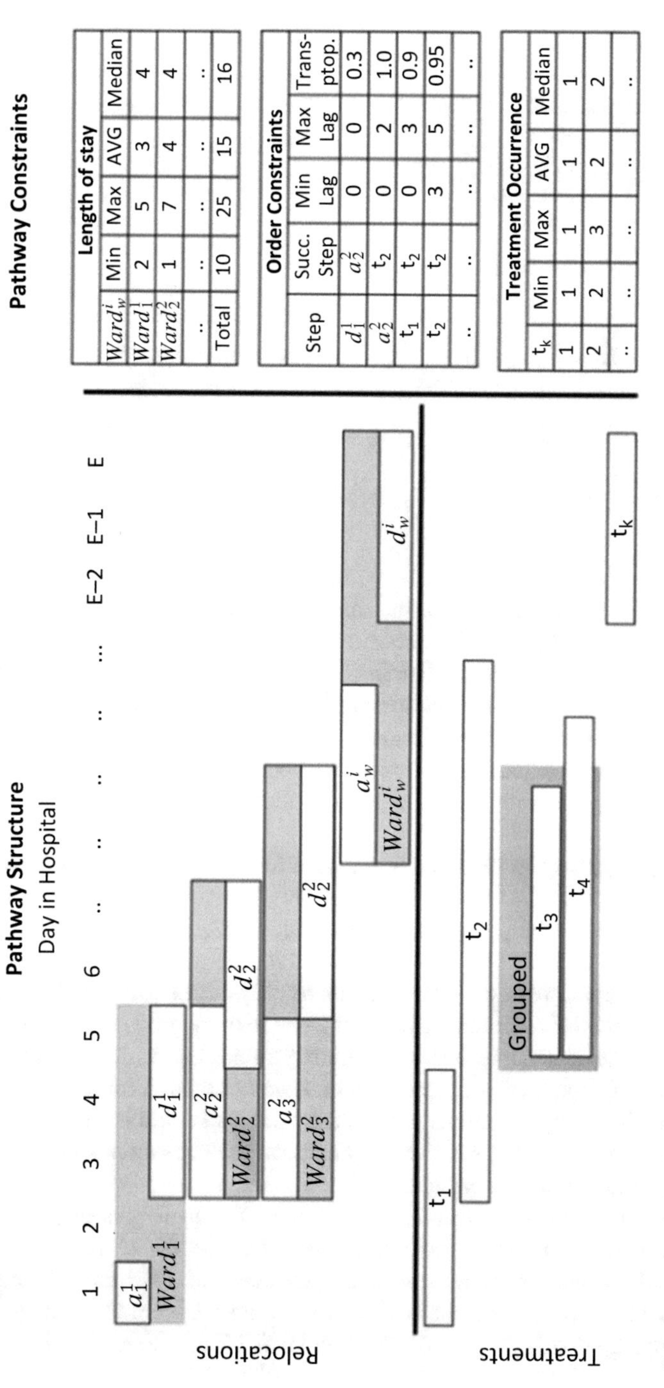

Fig. 2.2 Structure of mined pathway structure and constraints [4]

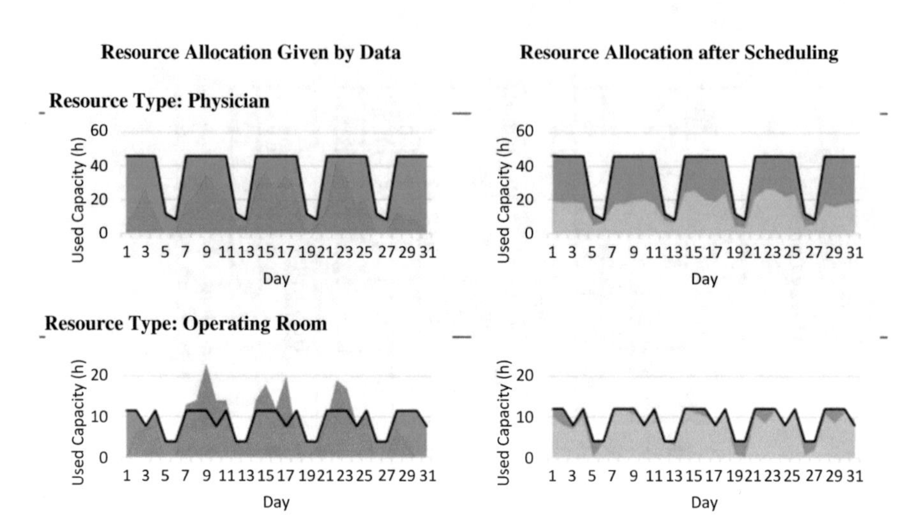

Fig. 2.3 Improved quality of planning results gained by prescriptive methods [5] using AI/OR synergy

In [5], it is shown that the information on pathway structure and pathway restrictions and rules gained by process mining as indicated in Fig. 2.2 can be effectively used in prescriptive planning of clinical pathways using mathematical optimization giving rise to a data-driven hierarchical mixed-integer programming approach. A real-world case of German university hospital department shows the benefits on the solution quality of prescriptive planning. The details can be inspected in [5]. Here, only some remarks about the quality of results gained by the AI/OR synergy are indicated.

Figure 2.3 depicts the resource usage given by (historic) data compared with the resource usage implied by the results of prescriptive optimization for the same period (as it were to be planned). For the resource type "physician," a smother resource usage is indicated by the optimization results without exceeding the given physician capacities. With the optimized plans, physicians may regularly integrate other tasks within their working days, giving rise to more productivity and planning stability. For the critical resource "operating room," a twofold quality benefit emerges: not only that the optimized plan indicates smother resource usage but also that the many capacity violations indicated by the data could be resolved. For this critical scarce resource, the solution of predictive methods attains higher resource utilization without violating capacities.

As our purpose in this paper is to discuss AI/OR synergies in a generic and abstract way, we consciously do not go here into details of how the pathway structure and constraints are integrated as variables and constraints of the mathematical model and we also do not dwell on the innovative hierarchical mathematical programming approach in [5]. At the top level of the hierarchical model, the planning of admissions for elective patients together with the scheduling of critical and complex treatments is formulated. In the top-level model, an anticipation of the base level is integrated. The base-level model includes the scheduling of other treatments together with ward and bed assignments. Recall that within hierarchical planning, top- and base-level models interact in terms of instruction and reaction [10]. One of the complex treatments considered in the higher hierarchical level are the operations in operating rooms. This further explains the good quality solutions indicated by a smooth and feasible solution with respect to the scarce resource "operating room" (the bottom part of Fig. 2.3).

This section had the purpose of introducing the types of synergies between descriptive/predictive and prescriptive analytics. Figure 2.1 showed two types of such synergies, where one of them is denominated by the 1. Type of AI/OR synergy. In this AI/OR synergy, complex structures are mined, as indicated in Fig. 2.2 for pathway structure and constraints, and then integrated into optimization models of prescriptive approaches. The quality of results should increase (as exemplarily shown in Fig. 2.3) because of the use of more accurate structures mined by AI methods. The next two sections will detect and discuss types of AI/OR synergies in a more generic and abstract way (Sect. 2.3). A new second type of AI/OR synergy will be detected and discussed in detail (Sect. 2.4).

2.3 Detecting AI/OR Synergies Within Hospital Decision Support: Interdependencies, Dimensions of Complexity, Two-Dimensional Scheme, and Types of AI/OR Synergy

As indicated earlier, effective decision support in hospitals depends on understanding the different decision-making tasks and their interrelations. Besides dealing with complex patient-centered multidisciplinary pathways, there are many hidden interdependencies, whose understanding heavily affects the effectiveness of decision support. We first discuss the different types of interdependencies apparent in complex decision-making environments like hospitals. Based on this discussion, two main dimensions of complexity are recognized: complex structures and complex environments. We then combine these two dimensions of complexity into a new kind of two-dimensional scheme. This scheme categorizes a magnitude of decision-making and decision support tasks in hospitals at a varying level of detail and consideration. By identifying AI and OR tasks within this scheme, many types of AI/OR synergies will be detected and discussed.

2.3.1 Types of Interdependencies: First Group

Interdependencies generally refer to interactions of at least two actors, processes or issues. In the current research literature on decision support in hospitals, these interdependencies are not examined and classified in detail. There are a few articles that discuss interdependencies in a hospital. Unfortunately they mostly focus on one kind of interdependencies. So a broad overview about interdependencies and its integration in planning models is lacking. Furthermore, in most cases researchers state that there are interdependencies and they make assumptions in order to cope with them. Because of that, this subsection provides a more detailed classification and analysis of possible interdependencies in a hospital. In general, it can be assumed that, depending on the level of abstraction of the decision problem, the nature and influence of interdependencies change. In many cases, this has a major impact on the modeling of the problem. The more the interrelationships are considered, the more complex the problem becomes [11]. In order to tackle their importance for decision-making, they are separated into two groups. The first group (Table 2.1) refers to the first synergy, which means that it is generally possible to integrate them into planning models. The second group (Table 2.2) is harder or even impossible to model and will be discussed in combination with a second type of synergy in Sect. 2.4.1.

The first type of interdependencies covers *capacity-based interdependencies*, which are characterized by competition for shared resources. For example, competition for CT scans. Basically different patients from different wards have a certain demand for scan resources. So these different patient pathways should be synchronized in order to achieve an evenly utilization rate without long waiting times. Depending on the level of abstraction, different subareas can be considered. On the one hand, there is competition between patients for beds and resources within

Table 2.1 First group of interdependencies in a hospital

No.	Interdependency	Properties
1	Capacity-based	Competition for shared resources
		A decision has a possible impact on further decisions
1	Staff-based	Employee specialization and hierarchy levels are considered
		Effects on staff planning and pooling
3	Problem-based	Relation between separate decision problems
		On the same or on different hierarchical levels

Table 2.2 Second group of interdependencies in a hospital

No.	Interdependency	Properties
1	Process-based	Effects of process improvements are investigated
		Effects of local optimization depend on the position in the system
2	Functional-based	Arising from treatment of various diagnoses
		With rising numbers of diagnoses uncertainties increase
3	Patient-based	Relationship between elective and emergency patients
		Forecasting of emergencies affects the available capacity for elective patients

a ward [12]. Since a bed can only be used by one patient at a certain time, it must be decided which patient is admitted to which bed. Moreover, there are different types of beds that are only suitable for different patient groups [13]. Additionally, capacity decisions on bed level have a huge impact on potential treatment capacities. Depending on the diagnosis, different treatments may be necessary. Therefore, a second competition arises. Certain diagnoses require treatment in specialized organizational units. The patients admitted compete for surgery timeslots or appointments in imaging departments. Decisions that are good at the bedside level, for example, optimizing the occupancy rate of beds, can have a negative impact on further capacity decision problems on treatment level. Therefore, the capacity bottleneck is pushed just one step further [13]. So optimizing resources for all patient groups at the same time is a complex task [12].

The next kind of interdependencies is composed of *staff interdependencies*. In a hospital, employees are highly specialized medical professionals. For example, nurses can have different specializations and hierarchy levels. In contrast, there is a possibility that medical professionals are cross-trained in different professions [13]. In addition to nurses, there may be head nurses with more responsibility and authority. Furthermore not all nurses are preassigned to a particular ward. For example, the concept of resource pooling is transferred to nurses in order to gain flexibility in nurse assignment in case of unforeseen events because they are distributed to different wards as needed. The interdependencies are particularly evident here as a result of personnel planning [16, 17]. If nurses with a special field of study are only employed at individual wards, this can lead to treatment or care bottlenecks at other wards [13]. Bottlenecks in treatment processes can appear if a nurse or a physician is forced to wait for the completion of another task. If there is no suitable employee ready for this task, delays appear inside the treatment process [13]. Therefore, a better consideration of staff-related interdependencies could lead to a better planning of staff-related questions of the treatment process [18]. One possibility of coping with these interdependencies is pooling of nurses. As required, these pooled nurses can be assigned to different wards according their qualification.

The last and mostly neglected but crucial kind of interdependencies within this group covers *problem-based interdependencies*. This kind describes the relation between separate decision tasks. In general, combinations of previous interdependencies are possible. For example, when a capacity planning of a ward is performed, the possible consequences for the planning of treatment capacities should be respected. This means that for a particular decision problem the related upstream or downstream problems should be respected in order to achieve results usable in practice. But in a hospital environment it is difficult to characterize upstream and downstream problems. In most cases a decision problem has multiple interdependencies with other decision areas that have multiple relations on their own.

In Fig. 2.4 two examples of multiple interdependencies are shown. On a strategical level, a connection between bed, treatment, and staff capacity is obvious. For example, with rising bed capacities in a ward, treatment and staff capacities should be adjusted as well in order to prevent additionally capacity bottlenecks.

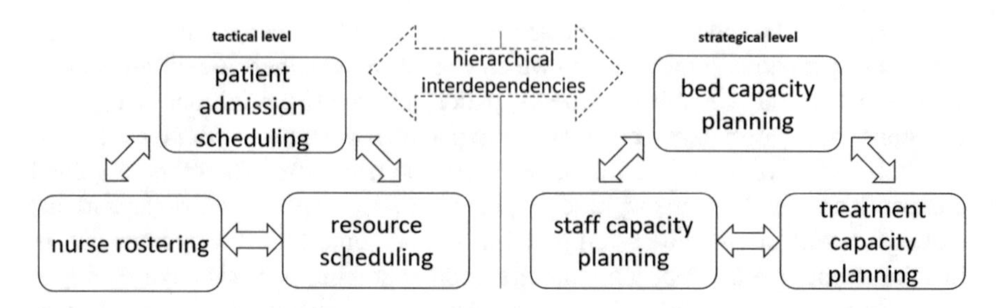

Fig. 2.4 Problem-based and hierarchical interdependencies

The same logic can be applied to the tactical level. Decisions made during the patient admission scheduling have a direct impact on decisions in other problem areas like nurse rostering or resource scheduling. Another form of problem-based interdependencies incorporates hierarchical interdependencies. That means that connections between decision problems at different decision levels are an important source of interdependencies. In Fig. 2.4 a hierarchical connection between bed capacity planning and patient admission scheduling is illustrated. If the bed capacity is high, the corresponding patient admission-scheduling problem is less complex. With a tighter planned bed capacity, the admission scheduling becomes more complex.

Despite of the inherent complexity of these interdependencies, it is possible to integrate them in planning models. In order to clarify their impact, two dimensions of complexity are introduced in the next subsection. Moreover, a two-dimensional scheme of decision tasks and the presented interdependencies are discussed as an approach to generate a first synergy between different decisions tasks.

2.3.2 Dimensions of Complexity and Overview About OR and AI Tasks and Synergies

Clinical pathways as complex structures represent a foundation of a lot of patient-centered planning problems, and their proper consideration is a general precondition for an effective hospital-wide decision support. But for an effective decision support, this is only one precondition. A complex environment has a huge impact on the complexity of planning problems. In general, patient pathways can consider one-to-many wards and different kind of resources. Also the environment creates another form of complexity within the decision support. Within the environment, there is a consideration of shared resources between multiple wards, general compliances that are present in a hospital, and other complex organizational conditions within a ward or the whole hospital. These different kinds of complexities, as shown in Fig. 2.5, are generating a lot of problems in the context of planning and decision

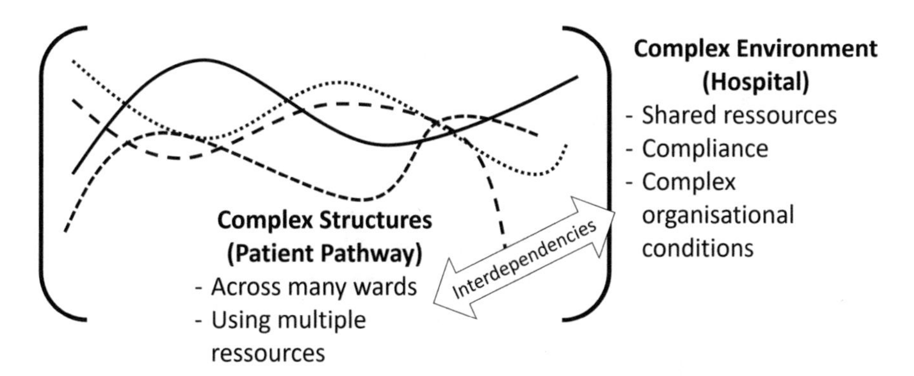

Fig. 2.5 Schema of confrontation of complex structures and complex environment in hospitals

support. While considering the combination of several dimensions of complexity, the discussion and investigation of different interdependencies in the last subsection is very important.

Because of the confrontation of two different kinds of complexities, a two-dimensional consideration of the problem is suitable in order to gain a deep understanding of the underlying quadratic complexity. In a consideration of complexity, the different dimensions can be investigated in different detail level, which have a great impact on the kind of decision support provided. The grade of complexity can be viewed from different starting points. First the complexity based on a consideration of patient pathways is shown. In the direction of the y-axis, the composition of a pathway is shown in different detail levels ranging from bed sequence to treatment sequences. On the x-axis, the multiplicity of pathways is considered. Here, an extension from abstract pathways to homogenous and further heterogenous pathways is illustrated. The most complex consideration is the observation of all possible pathways including irregularities during the pathway execution. The second possible point of view is the consideration of the resources of a patient pathway. Across the y-axis, the detailed level of resources needed by a pathway is shown, while the x-axis shows the extension of the scope of resources from one-to-many wards to a hospital-wide view. These considerations are illustrated in Fig. 2.6.

Within Fig. 2.6, the transition between strategical, tactical, and operational planning can be seen. Here, several grades of detail can be considered. It is important to notice that the complexity rises in two dimensions. From a consideration of single pathway and aggregated resources, a route to a considering all pathways and detailed resources has to be found. Moreover, with the help of this figure it is possible to link long-term, middle-term, and short-term decision support on an abstract level. This link is shown through the synergies of AI/OR methods on the left, bottom, and right side of the figure, which are discussed in detailed in the following (sub)sections.

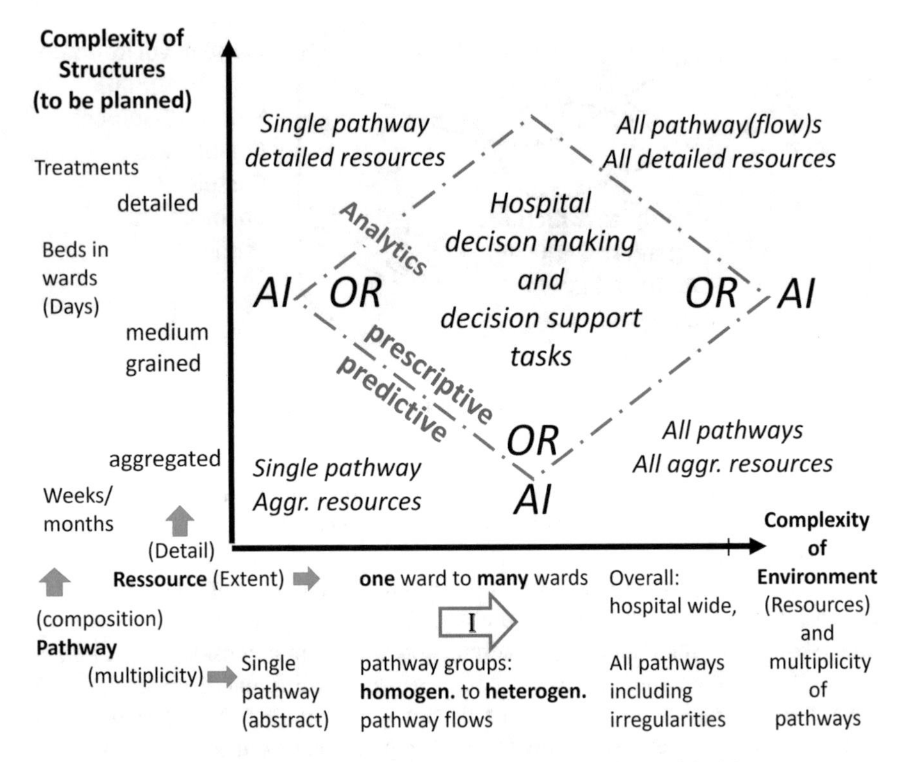

Fig. 2.6 Operationalization of the combination of complexities in hospital decision support

2.3.3 A New Two-Dimensional Scheme for Simulation-/Optimization-Based Decision Support in Hospitals Applied to Overall Bed Management in Interdependent Wards

A wide range of literature reviews show that a lot of problem areas inside a hospital exist. In [11], a general overview about using system sciences in health care is given. Based on that, some general problem areas are detected. In [19], different problem definitions regardless of their organizational unit are shown. Additionally two other surveys [20, 21] try to classify problem areas regarding their organizational scope of application. In addition to these general reviews, there are several specialized literature reviews that cover different scopes. This scope ranges from operating room scheduling [22] over capacity management [23] to capacity planning [24]. Furthermore, some reviews about different organizational units exist in the current research. There are special problem areas in intensive care [25], emergency departments[26, 28], or in specialized treatment areas like radiation therapy [27] or outpatient scheduling in general. It is obvious that there are several review articles. Nevertheless all presented review articles share the same problem.

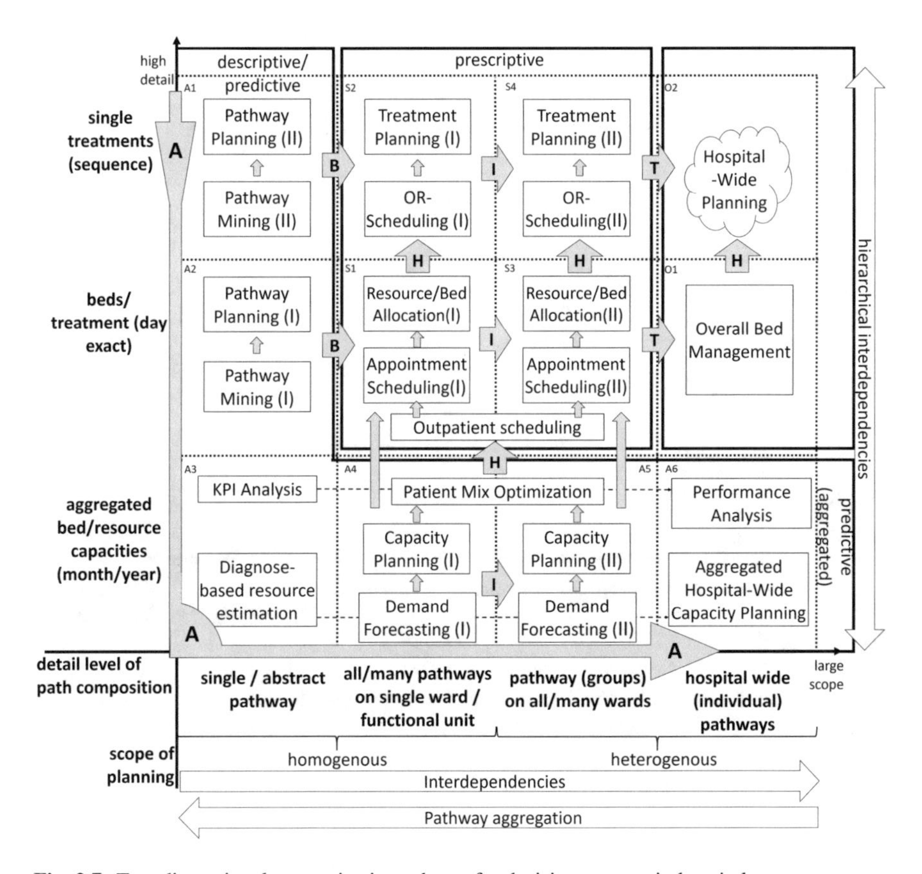

Fig. 2.7 Two-dimensional categorization scheme for decision support in hospitals

Based on the focus of the review, the existing categorizations are mostly based on problem areas or organizational units.

We propose a two-dimensional scheme depicted in Fig. 2.7 combining the complexity dimensions: complex structures (y-axis) and complex environments (x-axis), considered at different abstraction/aggregation levels. On the y-axis, the abstraction levels of resources are shown, which basically represents the inner composition of a patient pathway based on the granularity of resources needed. The x-axis shows different aggregation levels of patient pathways and pathway flows in the context of complex environments in a hospital. To clarify the discussion, only the amount of different pathways and their interaction with several departments inside a hospital are considered. By investigating the scheme, some general remarks about decision problems and interdependencies in a hospital environment are possible. In general, with an increasing path aggregation, the influence of interdependencies decreases. Nevertheless these interdependencies do not disappear. They are encoded into the aggregated pathways by making assumptions. Furthermore, an increasing pathway aggregation increases the homogeneity of pathways and patient groups.

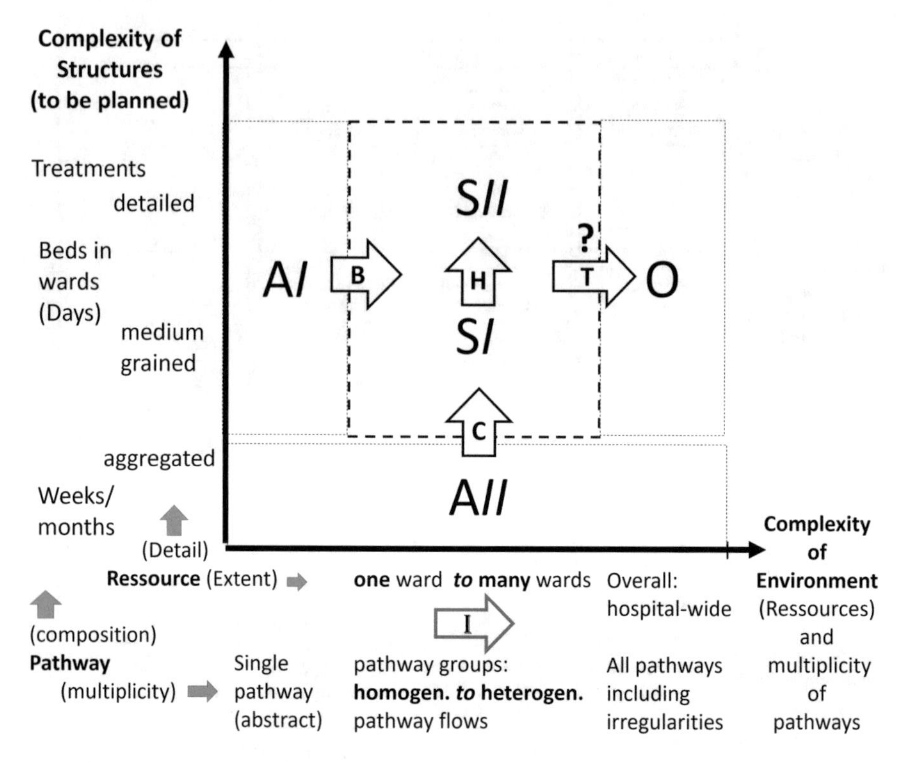

Fig. 2.8 Sectors of decision support in hospitals

This means that the corresponding decision problems do not have to deal with all possible interdependencies. In opposite direction, an increasing heterogeneity can be assumed, which leads to an increasing influence of interdependencies.

For the sake of clear derivation of types of AI/OR synergies, we will simplify the two-dimensional scheme of Fig.7 into that of Fig. 2.8 where only general problem sectors and their possible interactions are clearly highlighted. In discussing these different interactions between the sectors, the first type of AI/OR synergy discussed in Sect. 2.2 is shown and, furthermore, a hidden second type of AI/OR synergy is detected.

Considering the scheme, it is possible to derive several sectors, each one encompasses similar type of planning problems requiring the same special type of solving methods. These sectors are depicted in Fig. 2.8. Within sector AI, the task of pathway composition plays an important role. Within this sector, expert knowledge or process mining methods can be used in order to design patient pathways. Sector A II describes the problem area of strategical planning and performance analysis. Here, several tasks based on KPI and performance analysis can be done. In order to solve these tasks, a combination of data analysis methods, data mining approaches, and the aggregation of previously gathered AI results can be used. The scheduling-oriented sector S (divided into sector SI and SII) includes prescriptive planning tasks of groups of pathways (pathway flows). These groups of pathways can be homogenous if a single ward is covered and heterogenous if several wards are considered. Finally, the most complex sector of decision tasks in a hospital is illustrated by sector O (for overall and hospital-wide view). Within this sector, decision tasks are very complex because of the occurrence of many different pathways including irregular pathways that are not known in advance. So several new interdependencies arising inside this sector and this complex problematic will be resolved in Sect. 2.4 by the proposed AI/OR synergy of the second type.

The transitions B, C, H, and T indicate opportunities for different kinds of synergy. These synergy types will be discussed in the next subsections. For the transition T that completes the planning to hospital-wide or overall extent is linked to second type of AI/OR synergy and will be discussed in detail in Sect. 2.4. The question mark over the T-transition symbolizes that the operationalization of the associated AI/OR synergy of second type is neither straightforward nor similar to the other types of synergies including first type AI/OR synergy.

2.3.4 AI Tasks and AI/AI Synergy: Stepwise Aggregation from Process Mining to More Accurate Hospital Data Mining

One interesting synergy arises along arrow A in Fig. 2.7, which is detailed in Fig. 2.9. Here, a possible synergy between different AI methods is shown. Inside one dimension, there is an aggregation of resources to a single pathway for each diagnosis. In this part, one main method used is patient pathway mining. Basically, this kind of aggregation starts with medical needs on treatment level, where sequences of treatments are mined depending on diagnoses. With a rising level of aggregation, the focus of the mining changes. At the second level, the day-exact sequence of bed allocations is investigated and on the more strategical level general rules for capacity allocation are learned during the mining run in order to make a key performance indicator analysis or a diagnosed resource estimation. This also generates a starting point for an aggregation to groups of patient pathways. Within the first aggregation, several patient pathways at one ward are considered. It is possible to use standard data mining methods to make a demand forecast, which is enhanced during the usage of more aggregate environmental structures of the hospital. Generally, if the pathway is followed, it is possible to use sufficiently detailed input data and knowledge for each mining step. Furthermore the results of these approaches can be improved by using this new discovered synergy. If every step is done sequential, it can be assumed that a hospital-wide predictive planning uses detailed insights from a mining approach at treatment level. With a more detailed view on the input data, a more detailed result at a more aggregated level can be achieved.

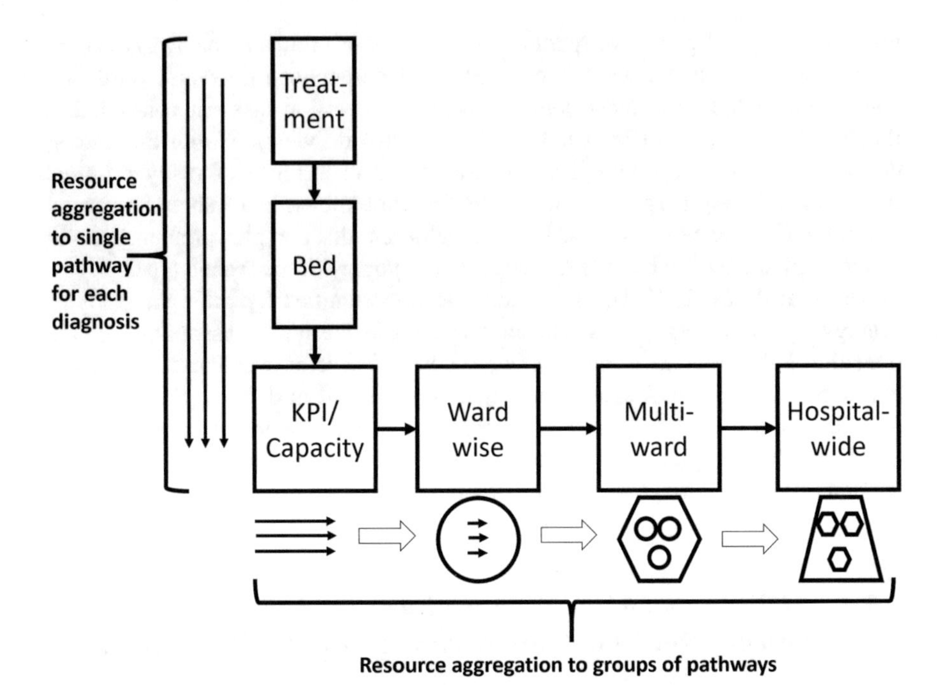

Fig. 2.9 Stepwise aggregation of process mining results

2.3.5 *OR Tasks and OR/OR Synergies*

Within Fig. 2.10, two different types of OR synergies are shown. Arrow H between the shown sectors describes a hierarchical synergy. The classical way of planning these different abstraction levels is the distinction between top and base levels (see [10]). In a standard consideration, the top-level problem deals with the planning of aggregated bed capacities and the base level describes the planning of single treatment sequences. Within the top level, an anticipation of the base level is done, which is clarified in another planning at base level. While considering basic resources, this type of hierarchical synergy is able to increase the quality of the results. But if the planning problem considers critical shared or scarce resource, this synergy could be regarded in different configuration as shown in Sect. 2.2.2 or in [5]. Within this configuration, the top level consists of planning scarce resources like operating rooms or main treatment resources for a certain patient pathway and an anticipation of a possible bed allocation is done. Afterward the exact planning of bed allocations and secondary treatments is done as a base-level problem. Another interesting interdependency arises along arrow I. Here, the transition from rather homogenous patient pathways for one ward to heterogenous patient pathways flows for a multiward setting is described. Basically, multicommodity (pathway) flow

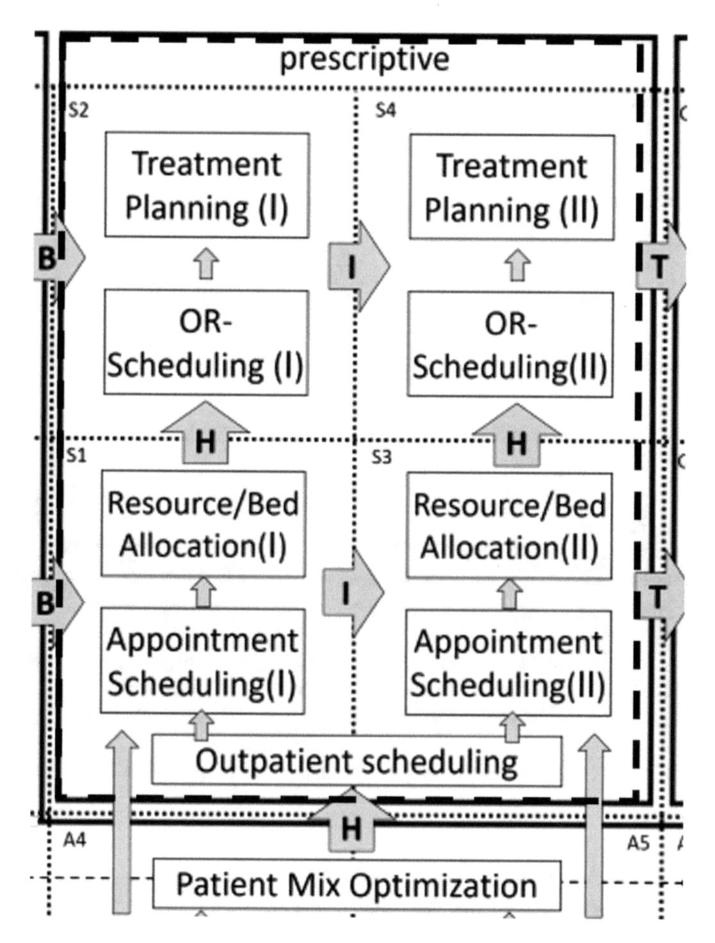

Fig. 2.10 S-Sector with OR tasks and OR/OR synergies

models are considered for modeling in order to increase the practicability of a planning and the efficiency of modeling.

2.3.6 *First Type of AI/OR Synergy and Detecting a Second Type*

The first type of synergy basically refers to Fig. 2.1 and the discussion in Sect. 2.2.2. Based on this rather economical discussion, a generalization considering the presented two-dimensional scheme is proposed. Referring to the aggregation level of pathway composition in the two-dimensional scheme, this type of AI/OR synergy could be viewed at different levels of detail.

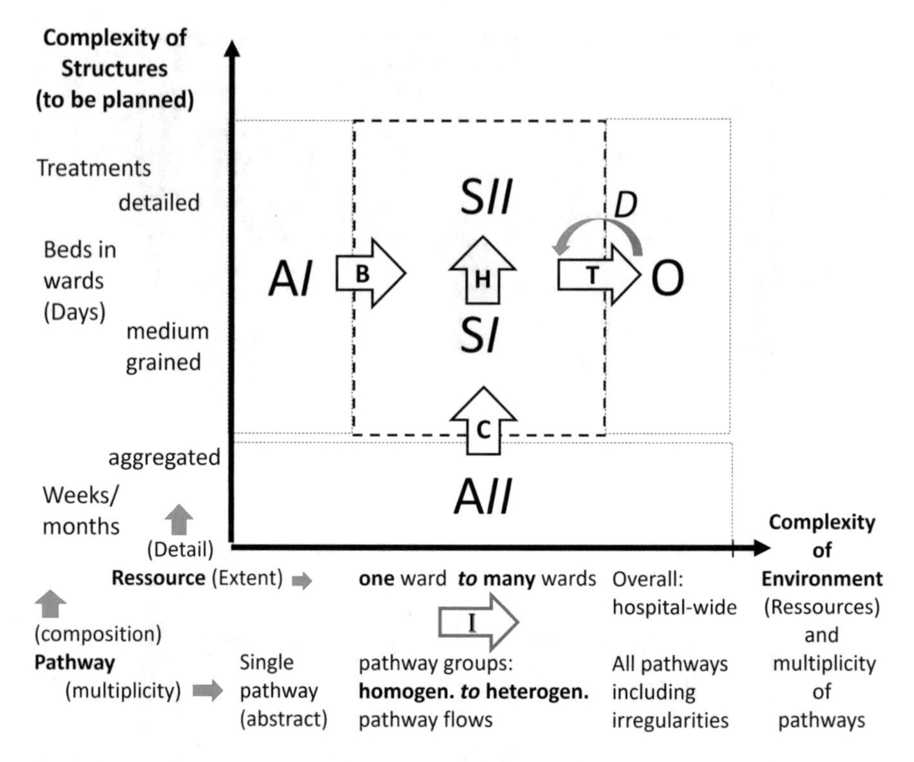

Fig. 2.11 Two-dimensional discussion of the combination of complex structures and environment

Generally, the first type of synergy describes the interplay between the descriptive/predictive sector A and the prescriptive sector S in Fig. 2.11. Generally, this synergy describes the usage of process mining or data mining methods as an input, for example as parameter or constraints, for prescriptive planning approaches. This kind of synergy can be divided into two subtypes. The first subtype deals with the possibility to learn details about the patient pathway composition via process mining as an input for an optimization (arrow B in Fig. 2.11). The second subtype is the usage of data mining methods in order to get good prognostic values about the path flow and its resource requirement.

Basically sector A describes a descriptive or predictive approach for strategical hospital planning tasks. These tasks present an input for sector S, which describe a prescriptive view of planning tasks. This input is shown as arrow B in Fig. 2.11. For a hospital, this is a crucial connection between strategical and tactical/operational planning areas. A high quality result of mining and aggregation procedures in sector A (A*I* or A*II*) is a mandatory requirement for a high quality planning result in sector S. This can be discussed with the example of pathway mining and planning and its implication for sector S. There is a differentiation into three different variants. The first variant is direct mining approach for patient pathways in the upper quadrant of sector AI. So the focus lies on details on treatments and their sequencing. A second variant is the aggregation of treatments on a daily level. The planned day of treatment inside a patient pathway is important (lower quadrant in sector A*I*). A third variant is a new combination of the previously discussed variants proposed in [4] and discussed earlier in Sect. 2.2. This variant aims at the utilization of pathway mining and planning for the more operational scheduling sector. In order to avoid a too narrow set of input parameters for operational planning, the exact location of each treatment inside a sequence is not important. Furthermore, the exact day of treatment is not important either. In order to provide decent input for planning within this variant, only mandatory relations between treatments and a range of possible treatment days are considered. So this variant provides an appropriate input for other planning tasks without excluding possible solutions by providing too strict input parameters.

This is one potential aspect of the usage of synergies between descriptive/predictive mining methods and prescriptive planning tools. A mining approach can be used in order to extract more general decision rules, which can be used as additional input parameter during the optimization. So the insights from a mining point of view are included. Nevertheless this possible combination cannot be extended for a hospital-wide view on decision support. Generally, this synergy can be described by mining relevant structures and constraints from patient pathways or patient data. These structures and constraints contain exigent knowledge that serves as an input for an optimization.

Within Fig. 2.11, several other kinds of synergies are evident. The consideration of a classical hierarchical synergy shown by arrow H between Sector SI and Sector SII is able to improve the practicability of the results. Furthermore, arrow C shows a generalization of classical synergies between predictive and prescriptive methods. So, for example, data mining of KPI structures and simple planning methods could be used in order to enhance the quality of prescriptive planning methods. Considering arrow T and its reverse arrow D, a new kind of synergy arises. While increasing the multiplicity of patient pathways and extending the scope of resource consideration, the complexity of the planning problems is huge. Clearly this synergy deals with the inclusion of irregularities inside the planning process. Within a complex organization like a hospital, not all patient pathways are known in advance and it is not possible to include all compliance and interdependencies arising for the whole hospital. In order to enhance the scope of planning, the nature and the occurrence of these irregularities are crucial for introducing a new area of decision support in hospitals. In fact, to integrate the transition T for a hospital-wide decision support, methods have to be implemented in order to integrate mined discrepancies in sector O within prescriptive planning methods in sector S (arrow D).

2.4 Second Type of AI/OR Synergy: Mining of Process Discrepancies and Its Interplay with Prescriptive Planning Toward Effective Hospital-Wide Decision Support

As indicated in Fig. 2.11, the second type of AI/OR synergy should help for the transition T from a multiward, multipathway flow optimization in sector SII toward a hospital-wide overall decision support (sector O). We first discuss a second group of interdependencies (Sect. 2.4.1), which may explain the model–reality gaps between solutions produced by prescriptive planning methods for the SII-sector and reality, that is when checking complex interrelationships not being modeled within prescriptive models. In order to operationalize the second type of AI/OR synergy (and the T-transition to the O-sector), we propose to introduce a mining procedure for process discrepancies that can be organized by type of interdependencies (Sect. 2.4.2). The backward D-arrow in Fig. 2.11 indicates the interplay between process discrepancy mining with prescriptive planning. This mechanism together with the operationalization of the second type of AI/OR synergy by a discrepancy-driven approach is discussed (Sect. 2.4.3).

2.4.1 Types of Interdependencies: Second Group and Model–Reality Gap

As stated before, we differentiate between two different groups of interdependencies. The second group consists of interdependencies that are very complex and hard to model in classical planning methods. In order to achieve a hospital-wide decision support, it is crucial to understand their behavior and try to incorporate them in planning approaches. Generally they refer to organizational structures. Thus they are enlarging the planning scope within the environment of the hospital. An overview about these interdependencies is given in Table 2.2.

The first type of interdependencies within this group involves the *process-based interdependencies*. These interrelations mainly relate to the effects of process improvements. A hospital can be seen as a so-called downstream system [14]. This means that the flow of patients is primarily directed in one direction. This means that patients enter the hospital via a few facilities, such as a central emergency room, and will be distributed to other facilities. A characteristic of downstream systems is the fact that an isolated view of a domain can have a negative impact on the overall performance of a system. The effects of local optimizations on the overall performance depend on the position of the considered area in the overall system. In particular, process improvements in the main admission facilities lead to problems in the following areas. This can be demonstrated by a simple example.

We consider a simple system consisting of a central emergency room and a downstream operating room with an attached intensive care unit. If the flow of an emergency department (ED) is significantly increased, it means that patients will need surgery capacities sooner. Thus, the capacity bottleneck no longer arises at the ED but in the operating room area. As a result, the emergency room and regular ward patients are given a longer waiting time for their surgery and the doctor's workload may be stretched beyond their limit. Thus, considering process improvements, the implications for neighboring or downstream facilities should always be included in the considerations [14].

Another type of interrelations consists of the so-called *functional interdependencies*. At this type, the orientation of the functional unit determines the influence of interdependencies. In principle, functional interdependencies denote the interdependencies that arise from the treatment of various diagnoses or diseases. With a rising number of diseases treated, the uncertainty for the planning process and for capacity supply also increases. Functional units that treat only one disease have little functional interdependence. For such units, standard processes can be developed in order to simplify planning, while units that treat a high number of illnesses need integrated planning solutions to incorporate the interrelationships. Thus, interdependencies within the functional unit have a much higher impact. For an overview about functional interdependencies, see Fig. 2.12 [15].

In connection with the last paragraph, a sixth kind of interdependency can be discussed. This kind covers *patient-based interdependencies* between elective patients and emergency patients. From a planning point of view, this is a crucial kind of interdependencies on the highest level of planning. The available capacity for elective patients depends on the quality of the forecast of emergency patients. As every planning task in a hospital is correlated to the amounts of patients, therefore it is crucial for hospital decision support to cope with this kind of interdependency.

In Fig. 2.13, a graphical interpretation of this central trade-off between elective and emergency patients is given. In general, there are two different strategies dealing with this relation. Basically a hospital has a fixed capacity, for example, the amount

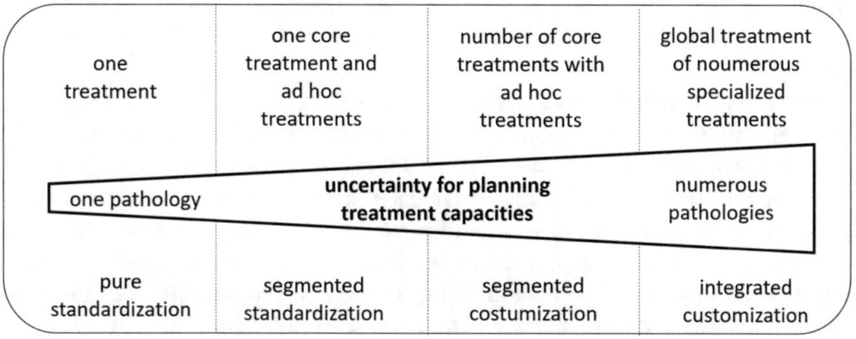

Fig. 2.12 Uncertainty through rising numbers of pathologies [15]

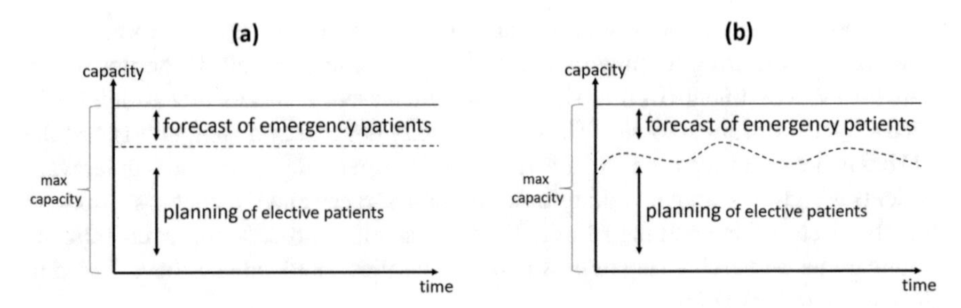

Fig. 2.13 Implications of patient-based interdependencies. (**a**) Fixed forecast strategy. (**b**) Flexible forecast strategy

of beds at a ward, for patient admission. The difficulty lies in the decision of how much capacity should be reserved for unplanned emergency patients. In Fig. 2.13a, a fixed strategy is presented. Based on a forecast, a fixed amount of beds over a longer period is reserved for emergency patients. The remaining amount can be used for planning elective patients. The performance of this strategy depends on the quality of the forecast of emergencies. If the forecasted amount is higher than the actual realized demand, the hospital has unused idle capacities. If the forecasted amount is lower than the realized demand, the hospital is overcrowded and patients have to be rejected. It is not allowed to reject emergency patients and therefore appointments of elective patients have to be canceled and in a worst-case scenario, all planning tasks located after an admission planning have to be redone. In order to cope with this problem, most hospitals reserve roughly 15% of its capacity for emergencies.

In Fig. 2.13b, a more flexible strategy is described. Here the amount of emergencies is forecasted for a shorter period of time or with different amounts. In general, a hospital is now able to react to a changing demand in emergency cases. In order to implement this strategy, a hospital is forced to adjust the planning of elective patients. So the planning of patients has to be done for shorter time periods in order to remain flexible in scheduling. So planning methods of a hospital have to be adjusted. Especially organizational changes in admission planning have to be ensured like short-term replanning of appointments of patients. In general, it could be possible that a hospital is able to use its capacity more efficiently with this strategy.

This group of interdependencies is complex in terms of modeling and integration into prescriptive planning models. Moreover not all aspects of the first group of interdependencies discussed in Sect. 2.3.1 can be modeled directly. Therefore model–reality gaps arise and harden the transfer of optimization results into practice. Because of this gap, a lot of results of planning problems cannot be transferred into practice. Within this gap, there are important organizational interdependencies that have an impact on the operational quality of planning results. For using an effective hospital-wide decision support, it is crucial to recognize these gaps and to make them as small as possible. Because of their complexity, new considerations regarding their integration into planning models have to be taken. One way to decrease the model–reality gaps is to use mining techniques in order to detect discrepancies during the patient pathways and gain knowledge from them for the future use within the planning process.

2.4.2 *Mining Process Discrepancies by Type of Interdependency*

In previous sections, the nature of patient pathways as complex structures, complex hospital environments, and the interdependencies between these two are discussed. In order to use these findings for an efficient decision support, a closer look at decision points inside a patient pathway is necessary. It is noticeable that these decision points harden the decision process even if only one patient pathway is considered. When considering multiple pathways across many wards, the complexity explodes. Fig. 2.14 shows a simple patient pathway with one decision point. After performing treatment A, a decision is made whether treatment B1 or B2 is performed. In fact the outcome of the decision depends on multiple factors like the result of the previous treatment, the properties of the patient or the capacity restrictions of the following treatments. Note that there are a lot more influencing factors for medical decisions during a patient pathway. Moreover, the previously discussed interdependencies have a huge impact on the decisions. That is the reason why these interdependencies have to be considered as detailed as possible within the decision process.

From a planning point of view, it is nearly impossible to optimize all decisions during a patient pathway or across multiple pathways. So a new planning approach has to be introduced. Normally a patient pathway has a standard pathway (given by the solid arcs in Fig. 2.14). In fact, the standard pathway is used for an initial planning for the patients. During the execution of the pathway, that means during the treatment process of a patient, the results of the treatment are stored. If a discrepancy against the standard pathway occurs, the medical expert has to make a decision on what alternative pathway should be taken by the patient. Here there are several possibilities. The expert could change to a well-known route through the pathway or decide to use another treatment that is not included within the patient pathway. For this online replanning, procedures can be implemented in order to decide which

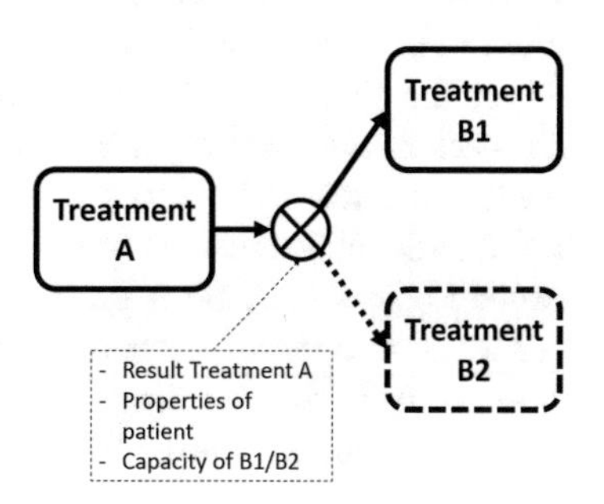

Fig. 2.14 Example of process discrepancies

alternative treatment or alternative department is suitable for the patient. In looking at discrepancies in a more general way over all pathways (mining methods for discrepancies), we may generalize the way of using alternative pathways in order to avoid these discrepancies and thus reduce the model–reality gaps.

Another way to look at process discrepancies is the change in process execution that has never happened before. On the first look, there is no possibility to react to such occurrences. For example, in the decision point shown in Fig. 2.14 there is no possibility to use treatment B1 or B2. So the medical expert has to come up with a unique solution. Another more practical example for the use of AI methods in this context would be at department level. If we assume that a patient can either be admitted to department 1 or 2 and none of them has any free capacity left. Then the case manager or medical expert has to find another department that is suitable for this patient in a short amount of time. In [3] the so-called ward cluster is used to find a suitable department or ward. This is a rather static approach to tackle this problem. In future research we propose the use of mining methods in order to achieve rules for dealing with such extreme situation. One solution could be the mining for similarities of wards. This could include the similarity in case of doing the same treatments or having similar qualified staff. If this is done on diagnosis level, this could lead to a set of breakout rules for strain situation in hospitals.

Generally from the clear differences between several kinds of interdependencies, there is a need for special mining techniques for each class of interdependencies. Ideally, the results of a mining process following these lines would resemble expert rules describing bottleneck situations in terms of general modeling entities, so that these learned rules could be integrated into prescriptive models.

2.4.3 Interplay Between Mining Process Discrepancies with Prescriptive Planning and Operationalization of the Second Type of AI/OR Synergy by a Discrepancy-Driven Approach

To illustrate the possible use of process discrepancies, three different approaches for manually detecting and using interdependencies and discrepancies for a more efficient decision support are shown. All of these are based on capacity evaluation at different abstraction levels.

For the first example, ward capacities are considered. In general simulation approaches, the ward capacity is considered as number of beds inside a ward. One result of our research [3] is a more detailed view on ward capacity. It was detected that interdependencies between male and female patients have a huge impact on available capacities. So a gender separation is an improvement of the modeling and leads to a more practical relevance of the results. A Second example in the same study is the handling of shortages of capacities at different wards. These strain situations are caused by interdependencies between several patient pathways on the same wards. So if a ward has no available capacity left, strategies for reallocation of patients have to be found. In [3] ward cluster is used to deal with this bottleneck in order to find a suitable ward, but this is a rather static approach. For an even more flexible strategy, the usage of similarity mining of wards could be encouraged in order to find suitable wards for patients. As a third example, the results of [5] are used. Here interdependencies between different treatments inside a patient pathway and across several pathways are considered. So for a scheduling of patients and treatments, these interdependencies are included inside the planning model. For this purpose, possible treatments are inserted into two groups. First complex treatments are scheduled, and afterward secondary treatments are planned. These three examples show the manual way to include extracted discrepancies into prescriptive planning models. The area of future research could be extended by searching ways to (partially) automate the inclusion of new interdependencies.

The inclusion of all possible rules inside an optimization model for planning purposes is nearly impossible. Some rules cannot be expressed and some others may result into optimization models which are prohibitively inefficient and thus not suitable for practical usage. That is why a new approach based on branch and cut is introduced, see, e.g., [29].

In general, branch and cut is an approach that enhances the well-known branch and bound algorithm by generating cuts as mathematical formulas inside the branch and bound tree. Within the presented approach in this chapter, interdependencies are seen as additional restrictions for a planning model. Because of the presented model–reality gap, it is not possible to exhaustively model all interdependencies within one planning model.

In Fig. 2.15, a new plan-and-refine framework for integrating new interdependencies into planning as a research agenda is proposed. This framework includes an

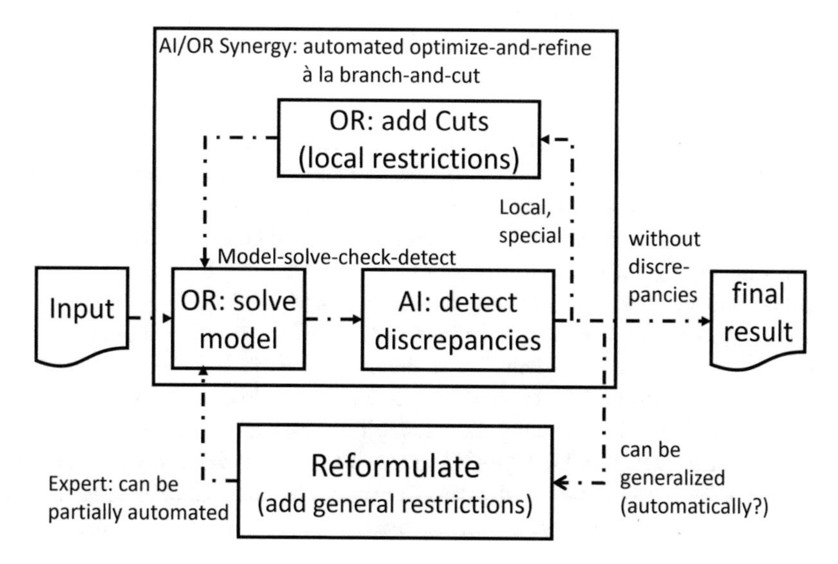

Fig. 2.15 Plan-and-refine framework including optimize-and-refine procedure

automated optimize-and-refine component operationalizing the AI/OR synergy of second type in the following form. First, the OR-based prescriptive model is solved, and then AI-mining procedure detects discrepancies (disagreement situations to reality). This is performed by a check routine incorporating compatibilities and compliance rules of hospitals. If discrepancies are found, some corresponding cuts (local restrictions avoiding the discrepancies) are added iteratively to the OR-based prescriptive model. In case a group of discrepancies of same type is recognized by the expert, he/she can reformulate the model by adding general restrictions. This model reformulation procedure is not done manually by the expert, but rather in a semiautomated way because the branch-and-cut kernel supports the expert by a listing of iteratively added constraints. This semi-automated procedure form together with the automated optimize-and-refine component the whole plan-and-refine framework.

2.5 Conclusion

In this chapter, we detected opportunities of synergy between artificial intelligence and operations research in the context of effective hospital-wide decision support. We showed that AI/OR synergy exhibits many facets and can take different forms. Observing that effective decision support in hospitals relies on considering complex entities to be planned in the form of patient-centered clinical pathway, a first type of synergy is discussed and its benefits considering real-world case studies are shown. Mining the structure and constraints related to the complex pathways helps integrating details on these objects within prescriptive mathematical optimization models.

By a more fundamental discussion around types of interdependencies, dimensions of complexity related not only to entities to be planned but also to resource environment, as well as of a two-dimensional characterization scheme for decision-making and decision support tasks in hospital, another hidden second type of AI/OR synergy is detected and its possible operationalization is discussed. We observed that although some types of interdependencies could be integrated in sophisticated optimization models for prescriptive planning, some other types of interdependencies in hospital decision-making could not be modeled. Thus, model–reality gaps emerge when trying to build prescriptive model for effective overall hospital-wide decision support. For the second type of AI/OR synergy, not the complex entities, but the complex hidden relationships when planning complex entities in complex environments are of interest. To gain these relationships, we proposed to mine process discrepancies (model–reality disagreements) in order to detect the relationships not being modeled within the prescriptive model. If these detected relationships exhibit a general form, an OR expert can integrate them into the model. Otherwise, the located discrepancies can be automatically used in order to generate cuts into the model, that is, restrictions prohibiting the same process discrepancies to occur in next optimization round. This is much like a branch-and-cut approach integrating an AI-oriented discrepancy mining procedure.

The second type of AI/OR Synergy will be the subject of future research where several questions should be answered. The first question is how to mine process discrepancies in an efficient and effective way. Here, detected types of interdependencies in this paper may guide us in formulating a shape-oriented inspection of model–reality gaps. The second question is how to automate or semiautomate a model-and-refine framework. Beginning with a branch-and-cut approach, it is necessary also to automate the generation of local restrictions (cuts) out of mined discrepancies. Last but not least, it is desirable then to automate the generalization of global restrictions out of many located restrictions (cuts) of the same type. This could be another form of AI/OR synergy: AI mining is here out of local restrictions (restriction mining) in order to gain general restrictions dealing with the same type of interdependency. The idea of learning association rules could be of help in order to learn not only general but also situation-conditioned restrictions.

References

1. Stoeck, T., Mellouli, T.: A two-dimensional categorization scheme for simulation/optimization based decision support in hospitals applied to overall bed management in interdependent wards under flexibility. In: Masmoudi, M., Jarboui, B., Siarry, P. (eds.) Operations research and simulation in healthcare, Springer (2020)
2. Helbig, K., Mellouli, T., Stoeck, T., Gragert, M., Jahn, P.: Simulation stationsübergreifender Patientenflüsse zur Evaluation flexibler Bettenbelegungsszenarien aufgrund der Jahresdatenanalyse eines Universitätsklinikums. In: MKWI 2014 – Multikonferenz der Wirtschaftsinformatik: 26. – 28. February 2014 in Paderborn: Tagungsband, 749–762. University of Paderborn (2014)
3. Helbig, K., Stoeck, T., Mellouli, T.: A Generic Simulation-Based DSS for Evaluating Flexible Ward Clusters in Hospital Occupancy Management. In: IEEE (eds.) Proceedings of the 48th Annual Hawaii International Conference on System Sciences, pp. 2923–2932 (2015)
4. Helbig [Schwarz], K., Römer, M., Mellouli. T.: A Clinical Pathway Mining Approach to Enable Scheduling of Hospital Relocations and Treatment Services. In Business Process Management, ed. Hamid Reza Motahari-Nezhad, Jan Recker, and Matthias Weidlich, 9253, pp242–250. Cham: Springer International Publishing (2015)
5. Schwarz, K., Römer M., Mellouli T.: A Data-Driven Hierarchical MILP Approach for Scheduling Clinical Pathways: A Real-World Study from a German University Hospital To appear in BUSINES RESEARCH (2016)
6. Salfeld, R., Hehner, S. P., Wichels, R.: Modernes Krankenhausmanagement: Konzepte und Lösungen. Springer (2008)
7. Roeder, N., Küttner, T.: Behandlungspfade im Licht von Kosteneffekten im Rahmen des DRG-Systems. Der Internist **47/7**, (2006)
8. Frese, E., Heberer, M., Hurlebaus, T., Lehmann, P.: Diagnosis Related Groups (DRG) und kosteneffiziente Steuerungssysteme im Krankenhaus. Schmalenbachs Zeitschrift für betriebswirtschaftliche Forschung **56/8**, 737–759 (2004)
9. Reinhold, T., Thierfelder, K., Müller-Riemenschneider, F., Willich, S.: Gesundheitsökonomische Auswirkungen der DRG-Einführung in Deutschland - eine systematische Übersicht. Das Gesundheitswesen **71/5**, 306–312, (2009)

10. Schneeweiß, C., Distributed Decision Making. Springer, Berlin (2003)
11. Atkinson, J., Wells, R., Page, A., Dominello, A., Haines, M., Wilson, A.: Applications of system dynamics modelling to support health policy. Public Health Research & Practice **25/3**, (2015)
12. Bakker, M., Tsui, K.: Dynamic resource allocation for efficient patient scheduling: A data-driven approach. Journal of Systems Science and Systems Engineering **26/4**, 448–462 (2017)
13. Green, L. V.: Capacity Planning and Management in Hospitals. In: Brandeau, M. L., Sainfort, F., Pierskalla, W. P. (eds.) Operations Research and Health Care, pp. 15–41. Kluwer Academic Publishers, Boston (2005)
14. Kolker, A.: Interdependency of Hospital Departments and Hospital – Wide Patient Flows. In: Hall, R. (eds.) Patient Flow, pp. 43–63. Springer, Boston (2013)
15. Lamothe, L., Dufour, Y.: Systems of interdependency and core orchestrating themes at health care unit level: A configurational approach. Public Management Review **9/1**, 67–85 (2007)
16. Burke, E. K., Curtois, T., Qu, R., Vanden Berghe, G.: A scatter search methodology for the nurse rostering problem. Journal of the Operational Research Society **61/11**, 1667–1679 (2010)
17. Burke, E. K., De Causmaecker, P, Berghe, G. V., Van Landeghem, H.: The State of the Art of Nurse Rostering. Journal of Scheduling **7/6**, 441–449 (2004)
18. Roche, K. T., Rivera, D. E., Cochran, J. K.: A control engineering framework for managing whole hospital occupancy. Mathematical and Computer Modelling **55/3–4**, 1401–1417 (2012)
19. Rais, A.,Viana, A.: Operations Rese3arch in Healthcare: a survey. International Transactions in Operational Research **18/1**, 1–31 (2011)
20. Fone, D., Hollinghurst, S., Temple, M., Round, A., Lester, N., Weightman, A., Roberts, K., Coyle, E., Bevan, G., Palmer, S.: Systematic review of the use and value of computer simulation modelling in population health and health care delivery. Journal of Public Health **25/4**, 325–335 (2003)
21. Günal, M. M., Pidd, M.: Discrete event simulation for performance modelling in health care: a review of the literature. Journal of Simulation **4/1**, 42–51 (2010)
22. Cardoen, B., Demeulemeester, E., Beliën, J.: Operating room planning and scheduling: A literature review. European Journal of Operational Research **201/3**, 921–932 (2010)
23. Jack, E. P., Powers, T. L.: A review and synthesis of demand management, capacity management and performance in health-care services. International Journal of Management Reviews **11/2**, 149–174 (2009)
24. Baru, R. A., Cudney, E. A., Guardiola, I. G., Warner, D. L., Phillips, R. E.: Systematic Review of Operations Research and Simulation Methods for Bed Management. Proceedings of the 2015 Industrial and Systems Engineering Research Conference (2015)
25. Bai, J., Fügener, A., Schoenfelder, J., Brunner, J. O.: Operations research in intensive care unit management: a literature review. Health Care Management Science **21/1**, 1–24 (2018)
26. Saghafian, S., Austin, G., Traub, S. J.: Operations research/management contributions to emergency department patient flow optimization: Review and research prospects. IIE Transactions on Healthcare Systems Engineering **5/2**, 101–123 (2015)
27. Vieira, B., Hans, E. W., van Vliet-Vroegindeweij, C., van de Kamer, J., van Harten, W.: Operations research for resource planning and -use in radiotherapy: a literature review. BMC Medical Informatics and Decision Making **16/1**, (2016)
28. Gul, M., Guneri, A. F.: A comprehensive review of emergency department simulation applications for normal and disaster conditions. Computers & Industrial Engineering **83**, 327–344 (2015)
29. Nemhauser, G., Wolsey, L.: Integer and Combinatorial Optimization. John Wiley & Sons, Inc. (1988)

Chapter 3
Real-Time Capacity Management and Patient Flow Optimization in Hospitals Using AI Methods

Jyoti R. Munavalli, Henri J. Boersma, Shyam Vasudeva Rao, and G. G. van Merode

Abstract Hospital systems are under constant pressure to provide quality care despite limited resources. However, traditional capacity management in hospitals is often not effective enough. One reason for this is the variability and uncertainty in the healthcare field that has to be managed. Another reason is the observation that hospitals are *open loop systems*, meaning they do not use feedback to determine if their output has achieved the desired goal of input. They do not observe the output of their processes controlled by them and use this information to take action. In hospital systems, there are few efficient planning systems or decision support systems to help administrators take decisions. This is different in other industries, where complex planning systems with the help of *Artificial Intelligence*, or AI as it is referred to, are often being used. This research chapter analyses the issues and possibilities for hospitals to incorporate AI into their capacity management and become intelligent systems in which operations and processes are regulated by feedback (*closed loop system*) and, more specifically, discusses the recent research of the authors on this topic, where Artificial Intelligent (Multi-Agent System) methods in combination with real-time coordination were described and implemented in the Aravind Eye Hospital (AEH) in India.

J. R. Munavalli (✉)
BNM Institute of Technology (VTU), Bangalore, India

Maastricht University, Maastricht, Netherlands
e-mail: jyotirmunavalli@bnmit.in

H. J. Boersma · G. G. van Merode
MUMC, Maastricht, Netherlands
e-mail: henri.boersma@mumc.nl; g.van.merode@mumc.nl

S. V. Rao
Forus Health, Bangalore, India

MUMC, Maastricht, Netherlands

M. Masmoudi et al. (eds.), *Artificial Intelligence and Data Mining in Healthcare*,
https://doi.org/10.1007/978-3-030-45240-7_3

3.1 Introduction

Due to growing patient demand, greater patient expectations and increasingly complex patient flow, hospital systems are under constant pressure to provide quality care despite limited resources [1–3]. Most of the time, capacity management in hospitals is not effective enough. This leads to a decreased accessibility of care and an increase in waiting times for patients resulting in reduced patient satisfaction and quality of care on the one hand, and low resource utilization on the other hand. One reason for this is the fact that planning is difficult because hospitals are influenced by their environment and experience variability and uncertainty due to unpredictable patient arrivals, varied service times and complex clinical pathways [4]. Another reason is however the observation that a lot of hospitals still are *open loop systems*, meaning they do not use feedback to determine if their output has achieved the desired goal of input. They do not observe the output of their processes controlled by them and use this information to take action. In hospital systems, there are few efficient planning systems or decision support systems to help administrators take decisions. This is different in other industries, where complex planning systems with the help of Artificial Intelligence, or AI as it is referred to, are often being used.

The research question for this chapter is therefore: Can an intelligent real-time management system using AI improve the hospital performance and transform the system to a control system in which operations and processes are regulated by feedback (*closed loop system*)?

This chapter analyses the issues and possibilities for hospitals to incorporate AI and become intelligent, closed loop systems. Also, research by the authors on real-time coordination technique are described to show the impact of AI on operations management. To understand the challenges for planning in hospitals, we will describe the general characteristics of operations management in hospitals.

3.2 Capacity Management in Hospitals

3.2.1 Traditional Hospital Capacity Management

Resource planning is based on aspects like average patient demand, the resource scheduling performed ahead of time and local optimization. In most hospitals, capacity/resources are planned and managed through a simple deterministic approach using average demand and average service times [5]. Hospitals not only treat scheduled patients but also walk-in patients making the patient arrivals of mixed type. The ratios between scheduled and walk-in patients vary depending on the type of hospital and its location. For example, the Aravind Eye Hospital in India treats more than 2000 walk-in patients a day [6–15]. Often hospitals fail to incorporate or overlook this variability and uncertainty during planning [16]. Ideally, hospitals should match the patient demand with their resources (supply).

When the patient demand is greater than the supply, it leads to increase in patients' waiting times and reduces accessibility whereas vice versa leads to low resource utilization [4]. Patient flow is the movement of patients through a hospital for their diagnostics and treatment and is central to patient experience or patient satisfaction. Suboptimal capacity management disrupts the continuous patient flow that in turn affects timeliness and quality of care [17].

The long-term capacity in a hospital is usually determined at an aggregate level [18]. Generally, hospitals forecast demand based on experience, historical data or sometimes advanced analytics. However, most hospitals have a functional structure with dependent and interconnected departments that manage their individual queues. Planning, scheduling and controlling of all the resources/capacity and patients are performed at department level. There are many instances where resources are scheduled beforehand to optimize their utilization and later patients are scheduled as per the availability of capacity/resources. Departments make their own decisions related to patients and resources with limited knowledge, which is local control. Resources in all the departments are scheduled based on either availability of resources or average demand over a period of time (day, week, or month). Clinics with low demand plan resources up to bimonthly, whereas clinics with high demand plan monthly or weekly [19, 20]. Also, the patient demand varies throughout the day and between the days. Because there is so much variability on operational level due to this functional structure, long-term capacity planning based on incomplete demand information (aggregate demand) does not fully reflect reality and leads to a mismatch. Often hospitals view this mismatch as a capacity problem (shortage in capacity), whereas they are likely caused by poor capacity management [21]. Van Merode et al. suggest the use of short-term planning when demand is non-deterministic [22]. Planning and control approaches that are commonly used are inadequate as they are not demand driven and lack synchronization. Patient flow and capacity management are interrelated to each other. To achieve an efficient patient flow, hospitals should manage capacity optimally.

3.2.2 Queuing and Synchronization in Hospitals

As written in the previous paragraph, hospitals have a number of departments that manage their own queues. Due to local control in departments, there is a lack of synchronization between departments, resulting in reduced patient flow and lower resource utilization. To analyse this problem, we will use the hospital system described in Fig. 3.1 as a queuing network with departments, such as D1, D2 and D3, pathways and control.

Hospitals are usually modelled by G/G/n models (general distribution where arrival time and service time processes are both arbitrary with "n" number of servers, arrival λ by General Distribution and service μ by General Distribution). All queues are first-come first-served, with queue lengths limited by the waiting area. The arrival rates and service rates are $\lambda 1$, $\lambda 2$, $\lambda 3$ and $\mu 1$, $\mu 2$, $\mu 3$ respectively.

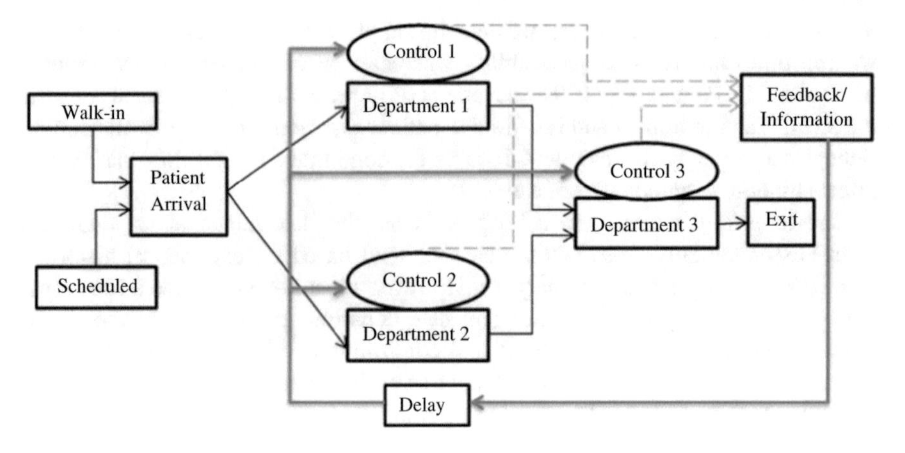

Fig. 3.1 Network representation of a hospital system

The patients in the hospital can follow any of the pathways, D1, D2, D1–D3 and D2–D3 depending on the type of care and patients' medical condition. Therefore, selection of the pathway is uncertain. New patient arrivals, along with the re-entry of patients from various departments, increase patient demand and result in variability. The schedules of resources or patients are based on forecasts of average demand. If the patient demand exceeds the planned number (due to walk-ins), the staff extend their working hours, resulting in staff overtime.

Ideally, to overcome the problem of synchronization, the manager in department 1 should use the average demand anticipation and negotiate for an extra resource during planning and scheduling, and by obtaining it, department 1 could treat the patients faster. The patients would be released from upstream department 1 to downstream department 3 without knowing the downstream department's status. This definitely improves the departmental efficiency and the waiting time of patients in department 1. However, department 3 receives patients faster from department 1 and is not ready to take many patients, as department 3 resources are limited. Therefore, department 3 has an increase in arrival rate but no change is caused in service rate, resulting in patients waiting longer in department 3. Department 1 improved its patient flow, but its efficiency does not reduce the patient's cycle time. Information about patient demand or available resources is not shared among the departments. This lack of synchronization between the departments creates bottlenecks in the hospital, disrupting the patient flow and causing an increase in latency.

The optimization of department 1 would increase the unregulated waiting in department 3. According to the schedule, department 3 could manage a certain level of threshold workload(t). When the workload in department 3 exceeds the threshold(t), the operational control 3 in department 3 identifies the difference between workload in queue and threshold workload and should signal department 1 and department 2 to block the entrance to these two departments. Therefore, according to the queue limit (waiting area) and resources (staff/doctors), the departments can set some rules based on which decisions can be made. Example: if (queue in department 3 is greater than fixed value t)→ (D1:↓ and D2 :↓). The hospitals often collect feedback after the event (end of the day) as shown in Fig. 3.1. Cause and effect are experienced during the same time, whereas action is taken later through planning and scheduling. The delay in responding to the change at hospital level (latency) can be either a day or a week or a month. Because of longer latency, the workflow cannot be optimized in the same timescale.

Currently, hospitals are incorporating data for their operations management [23] and applying technological solutions along with innovative approaches to improve the patient flow [24]. Data analytics has helped hospitals to analyse their patient flow accurately with appropriate data [25–27]. However, the latency of the data and information hinders the performance of the hospital system. The hospital system requires constant and real-time feedback from all subsystems, so that rescheduling is based on when and where the change is triggered. The delay in responding to change at department level (reaction time) depends on the flexibility of the subsystems in the hospital, whereas latency depends on the flexibility and synchronization in the hospital system. The hospital is an open-loop system that is affected by its surroundings, as there is a delay in utilizing the feedback and taking action to improve the workflow. It is this latency that categorizes the hospital as open loop system rather than closed loop system. To make hospital system from open loop to closed loop, the feedback should be used without delay. All the above methods of synchronization when performed manually are a cumbersome task with many complex tasks. Therefore, we will analyse how AI would fit into the above scenario to handle practical situations in real time.

3.3 AI Methods for Hospital Capacity Management

Artificial Intelligence, or AI as it is referred to, is the intelligence exhibited by machines. Recently, it has gained a lot of popularity with numerous applications in different industries. AI is also being used in healthcare, mainly in applications analysing the medical stages of diseases and the diagnosis of medical conditions. However, using AI for decision-making in planning and scheduling is still relatively new to healthcare (Fig. 3.2). Planning and decision-making are one of the most important aspects of operations management and AI powers advanced techniques and methods in this field, such as military decision-making [28], production planning and scheduling [29, 30], retail management, supply chain management [31] and communication network planning [32]. The planning and scheduling that involve intelligent decisions are

- Access management
- Queuing for workstations: waiting for doctors, nurses, eye screeners, etc.
- Patient throughput management

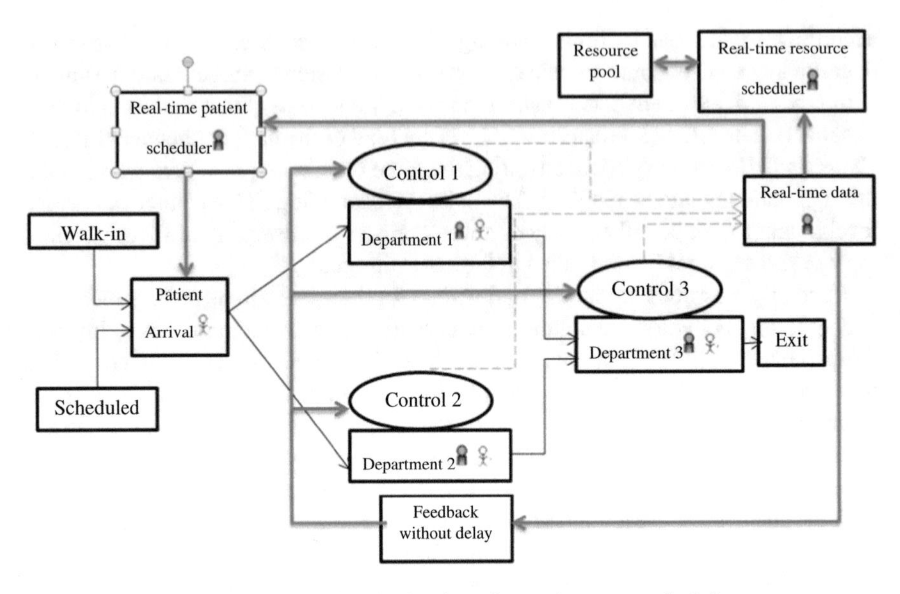

Fig. 3.2 Real-time control for synchronization in patient and resource scheduling

- Balancing workstations in patient pathways
- Resource utilization
- Demand-based scheduling
- Supply–demand match

Optimization models that are based on operations research and modern AI techniques could manage the hospital workflow through intelligent systems that are dynamic, robust and in real time. Various AI methods should be used to implement artificial intelligence, such as Multi-Agent System (MAS), Artificial Neural Networks (ANN) and Machine Learning (ML).

3.3.1 Multi-Agent Systems

MASs have been extensively applied to the complex healthcare environment [33–35]. Hospital systems consist of different types of people with different goals. It is observed that MAS are able to incorporate various attributes for modelling hospital systems. The MAS has been used in modelling emergency departments, operation theatres and inpatient hospitals [36, 37]. It has also been used for medical planning and diagnosis, where multiple strategies are analysed [38]. Patients and resources are described as agents that participate in scheduling [39, 40]. MAS is applied in coordinating resources that are shared across hospitals [41] and monitoring and forecasting patient demand [42]. Patients as agents are allowed to bid for the earliest treatment based on the resources auctioned [43]. MASs have been used in patient scheduling, where patients as agents compete for treatment appointments [39, 40, 43]. Schumann et al. [44] used a multi-agent approach to solve the conflict between local autonomy and global system performance. In their article, they show that the adjustment of autonomy of agents can lead to acceptable results retaining a certain level of overall system performance.

3.3.2 *Artificial Neural Networks (ANN)*

ANNs are biologically inspired computer programs that gather their knowledge by detecting patterns and relationships in data and learn, just as humans, through experience [45]. Deep learning algorithms are implemented in ANN for improved accuracy [46]. ANNs are used for forecasting patient length of stay, patient demand, patient treatment duration and patient appointment scheduling [47–49]. ANNs also have been used in improving capacity management, hospital efficiency, patient mortality and outcome of critically ill patients [45, 50–54].

3.3.3 *Machine Learning (ML)*

Machine learning is one of the methods of AI that helps the systems to automatically learn and improve experience and performance over a period of time. ML involves algorithms and mathematical models for learning process. Machine learning in medicine has recently made headlines. ML is used for tumour identification, disease development and diagnosis of diabetic retinopathy and many more [55–57]. Machine learning is used extensively in the field of data informatics.

3.4 Example of AI in Capacity Management and Patient Flow Optimization

In this section, we will describe one example of the AI methods for capacity management and optimization of patient flow implemented in a hospital. As described earlier with Fig. 3.1, manually gathering information about the process and optimizing flow throughout the day on every level of the organization is tedious and complicated through latency and a long reaction time. With AI methods, it is easier to manage and monitor information hospital-wide in real time. This way, all the departments become connected and synchronized. The negotiations for resources with traditional capacity management are a difficult task and involve a lot of communication. With the help of AI methods, these decisions can be made throughout the entire hospital, without delay, in real time.

3.4.1 Methods

For this example, the hospital in Fig. 3.1 was modelled as a Multi-Agent System where patients and resources like staff, doctors and nurses are agents that have their own goals with considerable amount of autonomy. By modelling hospital as a MAS, we incorporate a system's approach, making the workflow management an effective and efficient system [72]. Different types of agents are defined in the hospital:

- Patient agent (PA): These agents need care in the hospital. The database is maintained to track their movement in the departments according to clinical pathways.
- Resource agent (RA): Doctors, MRD staff and nurses are the RAs that perform the activities/tasks. The goal of these agents is to maximize their utilization.
- Department agent (DA): Keeps track of all patients and resources in the department.
- Route agent (RoA): This agent assesses the waiting time in each department.
- Patient scheduler agent (PSA): This agent creates RoAs who traverse the departments to find the optimal path for PAs.
- Resource scheduler agent (RSA): It monitors/tracks waiting time in all departments. It also monitors the resource pool and DAs and identifies the resource requirements in each department. This agent calls for bids from resources and schedules resources to reduce waiting time in the departments. With this agent, hospital can utilize its system status for scheduling both patients and resources.

The agents have attributes like name, address and identification number. An agent's class consists of agent's name, services and goals. The PA has database entries regarding the departments visited, in-time and out-time. Whenever a patient enters the hospital, the class application uses an interface agent and connects all other agents with the class application. The department, patient, resource and resource pool information are shared with the RSA. Now, the algorithms for patient scheduling, resource scheduling and real-time synchronization will result in a complete workflow optimization, as it is described as below (also shown in Fig. 3.2):

1. A resource pool keeps a record of all resources.
2. Real-time data is collected from all departments of the hospital.
3. Patients are scheduled to pathways based on the departmental system status.
4. Finally, resources are rescheduled using a synchronization mechanism.

All the participants in the hospital are mapped to agent roles in the MAS. Patients, resources in each department and managers are mapped as natural (human) agents (Fig. 3.2). Departments, patient scheduler, route agents (RoAs) and resource scheduler are defined to facilitate the necessary information and control operations and are mapped as artificial agents. Patients are passive agents, meaning the patients do not optimize their waiting time nor get involved in the optimization process. But, patients can also be active agents where they would choose their own pathways. A record for each PA is created when a patient enters the hospital. The database stores patient ID, age, in-time, departments visited and out-time.

With the hospital defined as Multi-Agent System, an Ant Colony Optimization algorithm can be used to increase hospital performance. An ant agent algorithm is extensively used in finding the shortest path and is used in transportation/logistic problems [58, 59]. Ants in the natural world wander randomly in search of food and leave a trail of pheromones while returning to their colonies after finding food. Other ants follow this trail of pheromones to find the food. Over a period, the shortest path will have the highest concentration of pheromones. However, over time the pheromone trail starts evaporating. This makes the ants search for alternative and sometimes better paths, thereby improving routing, congestion control, cost and time [60–62].

With certain modifications in this algorithm, it can be applied in flow optimization. The major difference between the logistics problem and hospitals is the sequential pathways in the workflow and stochastic pathways. The hybrid Ant agent algorithm is used to find optimal pathways for patients such that their waiting time is minimized. The algorithm uses the real-time status of all the patients, resources and departments to identify the total time a patient has to spend in the hospital [63–65]. In the MAS, at patient arrival (Fig. 3.2), the PA is scheduled to one of the pathways after the following actions are performed by the ant algorithm:

- The scheduler agent creates RoAs.
- RoAs gather waiting time information for all departments.
- The scheduler agent uses the collected information from RoAs and then chooses a pathway with the minimum waiting time and schedules the PA to that pathway. The ant agent algorithm was used for identifying an optimal path [60, 62].

To identify, the number of resources required as per the patient demand the term "Takt time" can be used. In the hospital context, Takt time is the average time at which a patient moves out of the hospital:

$$\text{Takt time} = \frac{\text{effective available time in a day}}{\text{no. of patients serviced in a day}}. \tag{3.1}$$

Takt time synchronizes demand and supply, commonly used in production industries [66–70]. For example, in Fig. 3.1, consider the department 1 with service time of 5 min and the patient arrival is 30 patients in an hour. In order to balance the patient flow, Takt time = 60 min/30 patients = 2 min/patient. The Takt time of 2 min does not mean that the patients are treated for only 2 min (contradicting the service times), but that every 2 min a patient should move out of the department. If the Takt time is less than 2 min, the service in the department is faster than patient demand and the resources either wait or stay idle. If it exceeds 2 min, then the patient waits. In order to achieve a Takt time of 2 min, the department needs resources:

$$r = \frac{\text{Service time}}{\text{Takt time}} = \frac{5}{2} \approx 3. \tag{3.2}$$

Demand-driven resource scheduling improves waiting time [70]. This example schedules resources that vary between different time slots (each time slot is of fixed amount of time). Here, the DAs collect the real-time data from their respective departments, and the required number of resources is identified. DA alerts RSA based on the threshold value for waiting times. The threshold value for each of the department varies as departments vary in service times. This threshold value is determined to avoid unregulated waiting times in departments. The existing resources and the required number of resources are compared. This difference is communicated, and the resources are rescheduled as per patient demand. The resource rescheduling will balance the patient flow throughout the departments of the hospital. It is referred to as synchronization of the departments in a hospital.

The additional resources (to match with demand) have to be communicated to all RAs. RSA uses auction-bidding, as it improves both hospital-wide (global) interest and self-interest of bidders (local). It removes the requirement of extensive one-to-one negotiations (time-consuming) between managers and resources. It enables comparison-based selection and fairly allocates resources to departments. The auction-bidding for resource scheduling through the n-player Bayesian game with incomplete information is appropriate. Here, in this case, the players (also bidders) are the RAs. The auction-bidding is a game between RAs, where they are competing with each other to achieve rewards by improving their utilization. The players have certain strategies and bid for the departments with slack resources.

The RSA initializes the auction by broadcasting the call for bids to all RAs in the units and resource pool. RA (bidder), $i = 1 \dots I$, observes the call and prepares his/her bid value v_i. All the bidders are interested in maximizing their rewards (utilization). There is no real cost associated, but the price that bidders offer is identified by the RSA. The bidding action shows the responsiveness of the RAs. RAs do not know about other RAs' bidding status. A set of auction rules or mechanism design will give rise to a game between the RAs. Bidders' information and value are independent (private) from each other. In this case, bidders submit their current utilization u_i (normalized value), time of last transfer t_i, the number of transfers till the time TR_i (at the start of the day it is 1) and the distance between current and required department d_i (stored in the database). With these auction rules, RAs play the game within the strategy space (all possible strategies/options). Bidders submit sealed bids b_1, b_2, ..., b_I. The bidder with the lowest value wins the bid (vacancy). The appropriate selection of the winner earns rewards for RSA. The goal of RSA is to maximize its rewards. The winner is transferred from the current department or from the resource pool to the required department, and the database is updated. The transfer of resources should be monitored. If the rescheduled RA does not reach the allotted department by a predefined time (in min), then the RSA again calls for bids. The above-mentioned model has been implemented in the Aravind Eye Hospital (AEH) that receives on average 2300 patients/day and has an OPC with 14 departments. AEH has resemblance to an assembly line system (a line of workers which a product being assembled passes consecutively from one operation to another until it is completed) [9, 14]. All patient arrivals are random (no appointment systems are used) and independent. This makes patient demand highly variable and uncertain. It is department centric and determines the resources much ahead of time based on average demand. The hospital has no control on input, as it provides the same-day care for all patients. The hospital consists of different types of people having different goals and has a hierarchical decision-making approach. Different people/staff play games to sub-optimally achieve their individual goals with respect to OPC-wide performance. The waiting times are improved by 56% (from 66.3 ± 18.7 to 32.1 ± 9.0 min) and resource utilization by 8.3% (from $78.7 \pm 2.6\%$ to $87.3 \pm 2.4\%$).

3.5 Conclusion

To sum up, hospitals are open loop systems that are prone to variability and uncertainty. This chapter emphasizes the importance of the combination of operations management research and artificial intelligence in healthcare and shows that it can be effective in improving performance in hospitals, not only theoretically but also in practice, as is shown in recent research on modelling the hospital as a MAS in the Aravind Eye Hospital (AEH). When AI is combined with operations management of hospitals, they can become intelligent, closed loop systems, with improved (real-time) decision-making and better overall hospital performance. Over time hospitals can be modelled as artificial neural networks that learn patterns from the previous hospital operations management and schedule resources and patients effectively and efficiently through advanced machine learning techniques.

3.6 Future of AI in Patient Flow Optimization and Capacity Management

The widespread usage of AI in healthcare is still a few years away [71]. Many factors have contributed to this slow adoption rate, including availability of data, poor data quality, low interoperability, concerns about patient privacy and challenges integrating the new technology into existing workflows. The developments around big data seem to be accelerating the AI implementation.

References

1. X. M. Huang, Patient attitude towards waiting in an outpatient clinic and its applications. *Health Services Management Research: An official Journal of the Association of University Programs in Health Administration/HSMC, AUPHA*, 1994. 7(1): p. 2–8.
2. D. I. Pillay, R. J. Ghazali, N. H. Manaf, A. H. Abdullah, A. A. Bakar, F. Salikin, et al., Hospital waiting time: the forgotten premise of healthcare service delivery? *Int J Health Care Qual Assur*, 2011. 24(7): p. 506–22.

3. Zhu, Z., Heng, B. H., and Teow, K. L. Analysis of factors causing long patient waiting time and clinic overtime in outpatient clinics. *Journal of medical systems*, 2012, 36(2), 707–713.
4. Ivan B. Vermeulen, Sander M. Bohte, Sylvia G. Elkhuizen, Han Lameris, Piet J.M. Bakker and Han La Poutre, Adaptive resource allocation for efficient patient scheduling. *Artificial Intelligence in Medicine*, 2009. 46(1): p. 67–80.
5. Paul R. Harper, A Framework for Operational Modelling of Hospital Resources. *Health Care Management Science*, 2002. 5(3): p. 165–173.
6. K V Ramani, Dileep V Mavalankar and Dipti Govil, Strategic Issues and Challenges in Health Management. 2008, India: SAGE Publications. 227.
7. C.K. Prahalad, The Fortune at the Bottom of the Pyramid, *Revised and Updated 5th Anniversary Edition: Eradicating Poverty Through Profits.*' 2009, New Jersey: Wharton School Publishing. 432.
8. Activity-Report, Aravind Eye Care System, 2014–15: Madurai. p. 76.
9. Larry Brilliant and Girija Brilliant, Aravind: Partner and Social Science Innovator (Innovations Case Discussion: Aravind Eye Care System). *Innovations: Technology, Governance, Globalization*, 2007. 2(4): p. 50–2.
10. Mehta, P., and Shenoy, S. Infinite vision: how Aravind became the world's greatest business case for compassion. 2011, Berrett-Koehler Publishers.
11. Dr. Bhupinder Chaudhary, Dr. Ashwin G. Modi and Dr. Kalyan Reddy, Right To Sight: A Management Case Study on Aravind Eye Hospitals ZENITH, *International Journal of Multidisciplinary Research* 2012. 2(1): p. 447–57.
12. V. K. Rangan and R.D. Thulasiraj, Making Sight Affordable (Innovations Case Narrative: The Aravind Eye Care System). *Innovations: Technology, Governance, Globalization Fall*, 2007. 2(4): p. 35–49.
13. Michael Moesgaard Andersen and Flemming Poulfelt, *Beyond Strategy: The Impact of Next Generation Companies*, 2014, New York: Routledge.
14. G. Natchiar, R. D. Thulasiraj and R. Meenakshi Sundaram, Cataract surgery at Aravind Eye Hospitals: 1988–2008. *Community Eye Health*, 2008. 21(67): p. 40–2.
15. Jyoti R. Munavalli, Shyam Vasudeva Rao, A. Srinivasan, A. Srinivas and Frits Van Merode, The Optimization in Workflow Management: Ophthalmology. *Journal of Health Management*, 2016. 18(1): p. 21–30.
16. T. B. T. Nguyen, A. I. Sivakumar and S. C. Graves, A network flow approach for tactical resource planning in outpatient clinics. *Health Care Management Science*, 2015. 18(2): p. 124–36.
17. Cinnamon A. Dixon, Damien Punguyire, Melinda Mahabee-Gittens, Mona Ho and Christopher J. Lindsell, Patient flow analysis in resource-limited settings: a practical tutorial and case study. *Global health, science and practice*, 2015. 3(1): p. 126–134.
18. J.M.H. Vissers, J.W.M Bertrand and G. de Vries, A framework for production control in health care organizations. *Production Planning & Control: The Management of Operations*, 2001. 12(6): p. 591–604.
19. Bruce D. Mansdorf, Allocation of Resources for Ambulatory Care -A Staffing Model for Outpatient Clinics. *Public Health Reports*, 1975. 90(5): p. 393–401.
20. Lynne C. Yurko, Tammy L. Coffee, Jane Fusilero, Charles J. Yowler, Christopher P. Brandt and Richard B. Fratianne, Management of an Inpatient-Outpatient Clinic: An Eight-Year Review. *Journal of Burn Care& Research*, 2001. 22(3): p. 250–4.
21. MM Rouppe van der Voort, FG van Merode and BH Berden, Making sense of delays in outpatient specialty care: A system perspective. *Health Policy*, 2010. 97(1): p. 44–52.
22. Godefridus G. van Merode, Siebren Groothuis and Arie Hasman, Enterprise resource planning for hospitals. *International Journal of Medical Informatics*, 2004. 73(6): p. 493–501.
23. Sara A. Kreindler, The three paradoxes of patient flow: an explanatory case study. *BMC Health Services Research*, 2017. 17(1): p. 481.
24. Lovett, P. B., Illg, M. L., & Sweeney, B. E. (2016). A successful model for a comprehensive patient flow management center at an academic health system. *American Journal of Medical Quality*, 31(3), 246–255.

25. Md Saiful Islam, Md Mahmudul Hasan, Xiaoyi Wang, Hayley D. Germack and Md Noor-E-Alam, A Systematic Review on Healthcare Analytics: Application and Theoretical Perspective of Data Mining. *Healthcare (Basel, Switzerland)*, 2018. 6(2): p. 54.
26. Daniel M. Bean, Clive Stringer, Neeraj Beeknoo, James Teo and Richard J. B. Dobson, Network analysis of patient flow in two UK acute care hospitals identifies key sub-networks for A&E performance. *PloS one*, 2017. 12(10): p. e0185912-e0185912.
27. Boersma, H. J., Leung, T. I., Vanwersch, R., Heeren, E., & van Merode, G. G. Optimizing Care Processes with Operational Excellence & Process Mining. *In Fundamentals of Clinical Data Science*, 2019, pp. 181–192. Springer, Cham.
28. Rasch, R., Kott, A., and Forbus, K. D.. Incorporating AI into military decision making: an experiment. *IEEE Intelligent Systems*, 2003, 18(4), 18–26.
29. B. Rodríguez-Somoza, R. Galán and E. A. Puente, Production Scheduling Using AI Techniques. *IFAC Proceedings Volumes*, 1990. 23(3): p. 387–392.
30. J. Mula, R. Poler, J. P. García-Sabater and F. C. Lario, Models for production planning under uncertainty: A review. *International Journal of Production Economics*, 2006. 103(1): p. 271–285.
31. E. W. T. Ngai, S. Peng, Paul Alexander and Karen K. L. Moon, Decision support and intelligent systems in the textile and apparel supply chain: An academic review of research articles. *Expert Systems with Applications*, 2014. 41(1): p. 81–91.
32. Hanning Chen, Yunlong Zhu and Kunyuan Hu, Multi-colony bacteria foraging optimization with cell-to-cell communication for RFID network planning. *Applied Soft Computing*, 2010. 10(2): p. 539–547.
33. Paul Bogg, Ghassan Beydoun and Graham Low, When to Use a Multi-Agent Systems?, *In Pacific Rim International Conference on Multi-Agents* 2008, Springer Berlin Heidelberg. p. 98–108.
34. Hanen Jemal, Zied Kechaou, Mounir Ben Ayed and Adel M. Alimi, A Multi Agent System for Hospital Organization. *International Journal of Machine Learning and Computing*, 2015. 5(1): p. 51–6.
35. David Isern and Antonio Moreno, A Systematic Literature Review of Agents Applied in Healthcare. *Journal of Medical Systems*, 2015. 40(2): p. 43.
36. R. Schmidt, S. Geisler and C. Spreckelsen, Decision support for hospital bed management using adaptable individual length of stay estimations and shared resources. *BMC Med Inform Decision Making*, 2013. 13(3).
37. Paulussen, T. O., Zöller, A., Heinzl, A., Braubach, L., Pokahr, A., & Lamersdorf W., Patient scheduling under uncertainty. *In Symposium on Applied Computing: Proceedings of the 2004 ACM symposium on Applied computing*, 2004, Vol. 14, No. 17, pp. 309–310.
38. Nieves, J. C., Lindgren, H., & Cortés, U. Agent-based reasoning in medical planning and diagnosis combining multiple strategies. *International Journal on Artificial Intelligence Tools*, 2014 23(01), 1440004.
39. Paulussen, T. O., Zöller, A., Heinzl, A., Pokahr, A., Braubach, L., & Lamersdorf, W. Dynamic patient scheduling in hospitals. *Coordination and Agent Technology in Value Networks*. GITO, Berlin, 2004, 149–174.
40. Paulussen, T. O., Jennings, N. R., Decker, K. S., & Heinzl, A. Distributed patient scheduling in hospitals. *In Proceedings of the Eighteenth International Joint Conference on Artificial Intelligence (IJCAI-03)*. 2003. Morgan Kaufmann.
41. U. Deshpande, A. Gupta and A. Basu, A distributed hospital resource scheduling system using a multi-agent framework. *IETE Technical Review*, 2001. 18(4): p. 263–75.
42. Stiglic, G., & Kokol, P. Intelligent patient and nurse scheduling in ambulatory health care centers. *In 2005 IEEE Engineering in Medicine and Biology 27th Annual Conference*, 2006, pp. 5475–5478.
43. Zöller, A., Braubach, L., Pokahr, A., Rothlauf, F., Paulussen, T. O., Lamersdorf, W., & Heinzl, A. Evaluation of a Multi-Agent System for Hospital Patient Scheduling. *ITSSA*, 2006 1(4), 375–380.

44. R. Schumann, A. D. Lattner and I. J. Timm, Management-by-Exception - A Modern Approach to Managing Self-Organizing Systems. *Communications of SIWN*, 2008. 4: p. 168–172.
45. S. Agatonovic-Kustrin and R. Beresford, Basic concepts of artificial neural network (ANN) modeling and its application in pharmaceutical research. *Journal of Pharmaceutical and Biomedical Analysis*, 2000. 22(5): p. 717–727.
46. Schmidhuber J., Deep learning in neural networks: An overview. *Neural Networks*, 2015. 61: p. 85–117.
47. Gül, M., & Güneri, A. F, Forecasting Patient Length of Stay in an Emergency Department by Artificial Neural Networks. *Journal of Aeronautics and Space Technologies*, 2015. 8(2): p. 43–48.
48. Chang, W. J., & Chang, Y. H., Design of a Patient-Centered Appointment Scheduling with Artificial Neural Network and Discrete Event Simulation. *Journal of Service Science and Management*, 2018. 2018(11): p. 71–82.
49. Tsai, P. F. J., Chen, P. C., Chen, Y. Y., Song, H. Y., Lin, H. M., Lin, F. M., & Huang, Q. P. Length of hospital stay prediction at the admission stage for cardiology patients using artificial neural network. *Journal of healthcare engineering*, 2016, p. 11.
50. G. Hao, Kin Keung Lai and Manzhi Tan, A Neural Network Application in Personnel Scheduling. Vol. 128. 2004. 65–90.
51. Ali Guneri and Alev Gumus, The Usage of Artificial Neural Networks For Finite Capacity Planning. Vol. 15. 2008. 16–25.
52. El Adoly, A. A., Gheith, M., & Fors M. N., A new formulation and solution for the nurse scheduling problem: A case study in Egypt. *Alexandria Engineering Journal*, 2018. 57(4): p. 2289–2298.
53. Raghupathi, V., & Raghupathi, W., A Neural Network Analysis of Treatment Quality and Efficiency of Hospitals. J Health Med Informat, 2015.
54. Clermont Gilles, Derek C. Angus, Stephen M. DiRusso, Martin Griffin, Linde-Zwirble and Walter T, Predicting hospital mortality for patients in the intensive care unit: A comparison of artificial neural networks with logistic regression models. *Critical Care Medicine*, 2001. 29(2): p. 291–296.
55. D. A. Clifton, K. E. Niehaus, P. Charlton and G. W. Colopy, Health Informatics via Machine Learning for the Clinical Management of Patients. *Yearbook of medical informatics*, 2015. 10(1): p. 38–43.
56. Carson Lam, Darvin Yi, Margaret Guo and Tony Lindsey, Automated Detection of Diabetic Retinopathy using Deep Learning. AMIA Joint Summits on Translational Science proceedings. *AMIA Joint Summits on Translational Science*, 2018. 2017: p. 147–155.
57. Kourou, K., Exarchos, T. P., Exarchos, K. P., Karamouzis, M. V., & Fotiadis, D. I., Machine learning applications in cancer prognosis and prediction. *Computational and Structural Biotechnology Journal*, 2015. 13: p. 8–17.
58. Kanaga, E. G. M., Valarmathi, M. L., & Rose, J. D., Coordinated Multi-Agents Based Patient Scheduling Using Genetic Algorithm. *ACEEE Int. J. on Communication*, 2010. 03(02).
59. Du, G., Jiang, Z., Yao, Y., & Diao, X., Clinical Pathways Scheduling Using Hybrid Genetic Algorithm. *Journal of Medical Systems*, 2013. 37: p. 9945.
60. Dorigo, M., & Gambardella, L. M., Ant colony system: a cooperative learning approach to the traveling salesman problem. IEEE Transactions on Evolutionary Computation, 1997. 1(1): p. 53–66.
61. Bean, N., & Costa, A., An analytic modelling approach for network routing algorithms that use "ant-like" mobile agents. *Elsevier Computer Networks*, 2005. 49(2): p. 243–68.
62. Luca Maria Gambardella and Marco Dorigo, An Ant Colony System Hybridized with a New Local Search for the Sequential Ordering Problem. *INFORMS Journal on Computing*, 2000. 12(3): p. 237–55.
63. R. Dybowski, V. Gant, P. Weller and R. Chang, Prediction of outcome in critically ill patients using artificial neural network synthesised by genetic algorithm. The Lancet, 1996. 347(9009): p. 1146–1150.

64. Majid Noor and Vinay Narwal, Machine Learning Approaches in Cancer Detection and Diagnosis: Mini Review. 2017
65. Jyoti R Munavalli, Shyam Vasudeva Rao, Aravind Srinivasan and GG van Merode, Integral patient scheduling in outpatient clinics under demand uncertainty to minimize patient waiting times. *Health Informatics Journal*, 2019. 0(0): p. 1460458219832044.
66. J.K Liker, The Toyota Way: 14 Management Principles from the World's Greatest Manufacturer. 2004, New York: McGraw-Hill. 330.
67. Wallace J. Hopp and Mark L. Spearman, Factory Physics-Foundations of Manufacturing Management. 2001, Irwin: McGraw-Hill. 698.
68. M. Eswaramoorthi, G. R. Kathiresan, T. J. Jayasudhan, P. S. S. Prasad and P. V. Mohanram, Flow index based line balancing: a tool to improve the leanness of assembly line design. *International Journal of Production Research*, 2012. 50(12): p. 3345–58.
69. JamesMac Gregor Smith and Baris Tan, Handbook of Stochastic Models and Analysis of Manufacturing System Operations. *International Series in Operations Research and Management Science*. Vol. 192. 2013, New York: Springer-Verlag 373.
70. Jyoti R. Munavalli, Shyam Vasudeva Rao, Aravind Srinivasan, Usha Manjunath and G. G. van Merode, A Robust Predictive Resource Planning under Demand Uncertainty to Improve Waiting Times in Outpatient Clinics. *Journal of Health Management*, 2017. 19(4): p. 1–21.
71. Jianxing He, Sally L. Baxter, Jie Xu, Jiming Xu, Xingtao Zhou and Kang Zhang, The practical implementation of artificial intelligence technologies in medicine. *Nature Medicine*, 2019. 25(1): p. 30–36.
72. Jyoti R Munavalli, Shyam Vasudeva Rao, A. Srinivasan and G. G van Merode, An Intelligent Real-Time Scheduler for Out-Patient Clinics: A Multi-Agent System Model. *Health Informatics Journal*, 2020. http://dx.doi.org/10.1177/1460458220905380.

Chapter 4
How the Health-Care Expenditure Influences the Life Expectancy: Case Study on Russian Regions

Nenad Mladenovic, Olga Rusetskaya, Souhir Elleuch, and Bassem Jarboui

Abstract One can expect that the life expectancy of people in a city or geographical region depends on health-care infrastructure in that city or region, as well as on investment devoted to it. In this paper we wanted to check the influence of health-care supports of different kind on the life expectancy. Data are collected on all 85 geographical districts in Russia, covering 15-year period. The symbolic regression model is applied and solved by variable neighborhood programming, the recent promising automatic programming technique. In other words, the analytic function is searched to present relation between the life expectancy and a few selected health-care financial attributes. Some years are used as training set, and some as testing set. Interesting results are obtained and analyzed. They confirm the fact that symbolic regression and artificial intelligence techniques might be the right approach in estimating the life expectancy.

4.1 Introduction

Life expectancy is an indicator that reliably shows the general state of health in a society in a certain period of time. The public health model developed by Lalonde [6] recognizes that in addition to the health system, factors such as the environment, genetic characteristics, and behavior (determinants of health not related to the

N. Mladenovic (✉)
Research Center of Digital Supply Chain and Operations, Department of Industrial and Systems Engineering, Khalifa University, Abu Dhabi, UAE

O. Rusetskaya
Leontief Centre, Saint Petersburg, Russia
e-mail: olga@leontief.ru

S. Elleuch
Department of Management Information Systems, College of Business and Economics, Qassim University, Buraidah, Qassim, Saudi Arabia

B. Jarboui
Higher Colleges of Technology, Abu Dhabi, UAE

M. Masmoudi et al. (eds.), *Artificial Intelligence and Data Mining in Healthcare*,
https://doi.org/10.1007/978-3-030-45240-7_4

health system) affect the health of the population. The results of numerous studies using the approach of the production function of health and highlighting individual factors affecting the health of the population allow us to group these factors into large blocks: heredity, socioeconomic factors, lifestyle, environment, and health-care system. Factors affecting health are closely related. Modeling health indicators depending on individual factors determining it is a methodologically complex task.

The research topic we are considering in this paper appears to be important. At the "Our World in Data" site (https://ourworldindata.org/), it is included in the list of 297 world's largest problems [1]. The web page's purpose is expressed as *Research and interactive data visualizations to understand the world's largest problems*. There, the life expectancy y is considered as a function of one variable, i.e., health-care expenditure in each particular country. A dataset of almost all countries are also provided at the site. A simple one-dimensional plot is presented for each country, showing mostly linear dependencies.

In [2], authors analyze life expectancy of 70 years old people, as a function of their physical condition and health-care money spent until they die, using the 1992–1998 *Medicare Current Beneficiary Survey* in the USA. They found that persons with better health had a longer life expectancy than those with poorer health but had similar cumulative health-care expenditures until death.

To estimate life expectancy y, the input of the health-care system is expressed by health-care expenditures per capita (current US$) as a single attribute in [3]. The data are collected for 175 world countries, grouped according to the geographic position and income level, over 16 years (1995–2010). Authors applied a panel data analysis. The obtained results show a significant relationship between health expenditures and life expectancy. Country effects are significant and show the existence of important differences among the countries.

In 1995–2015, most OECD countries constantly increased their spending on health care, but to varying degrees and with different effects of increasing life expectancy [10]. This is well illustrated by the example of some countries with the highest level of GDP per capita (e.g., the USA, Germany, Great Britain, Japan, Italy, France, Canada, Norway, the Netherlands, and Australia) [11]. During this period, life expectancy at birth and the average per capita income for health care increased in all these countries, but not everywhere the level and increase in health expenditure were accompanied by an adequate increase in life expectancy. This may indicate that lifestyles and behavioral stereotypes play an important role in increasing life expectancy, as well as the effectiveness of the use of funds entering the health-care system. The existence of a direct relationship between life expectancy and the level of health expenditure is also established on the basis of available data for OECD countries and partner countries in the year 2015 (or close to it) [12]. The following was found by the World Health Organization in [13]: The higher the proportion of government spending on healthcare in 2014, the higher the value of life expectancy at birth in 2015, with exception of the African region.

There are many other studies regarding life expectancy of people. They compare life expectancy between men and women, different geographical regions, countries, continents, etc. However, most of them consider just one input variable separately in showing the results. In this paper we use three different input variables simultaneously, all three being a kind of health-care expenditure. Therefore, we use multidimensional regression analysis to find functional dependencies more precisely. Moreover, for the first time we applied an artificial intelligence approach in estimating the life expectancy. We applied variable neighborhood programming technique, a recent automatic programming method for solving symbolic regression problem. In addition, we collected the relevant data for all Russian geographical regions. Interesting results are derived from the final regression formulas. For example, some input expenditures are not relevant for the life expectancy, and also the functions are mostly linearly increasing.

In the next section we define the estimation of life expectancy as a symbolic regression problem. In Sect. 4.3 we give a brief explanation of how the variable neighborhood programming (VNP) is working in solving symbolic regression problem, while Sect. 4.4 provides details regarding a case study on Russian geographical regions. Results and their analysis are provided in Sect. 4.5, while Sect. 4.5 concludes the paper.

4.2 Life Expectancy as a Symbolic Regression Problem

To estimate life expectancy, we use artificial intelligence and machine learning approach. More precisely, for each geographical region, we are trying to find the best analytic function $y = f(x)$, where $x = (x_1, x_2, x_3) \in R^3$.

Since we have a machine learning problem, the dataset must be divided into learning and testing sets. Therefore, the method contains two steps:

- **Learning step:** The AI or machine learning algorithm is applied to the 2/3 of the dataset of each district and
- **Testing step:** The AI algorithm is applied to the remainder of the dataset of each district (see Fig. 4.1).

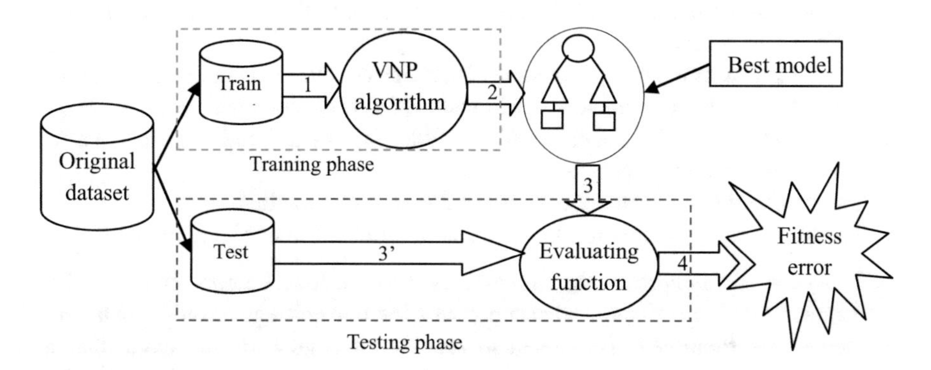

Fig. 4.1 Learning process schema

The symbolic regression problem consists of finding a mathematical relation in a symbolic form between inputs and the output. Our economically based input variables are

x_1 Expenditures of the consolidated budgets of constituent entities of the Russian Federation on health, million Roubles per 1 thousand people;

x_2 Expenditure of territorial funds of compulsory medical insurance of constituent entities of the Russian Federation, million roubles per 1 thousand people;

x_3 Average per capital money income of population, per month, in roubles.

We decided to eliminate the crude death as an attribute. Our only output variable y represents the life expectancy.

4.3 Variable Neighborhood Programming for Solving Symbolic Regression Problem

To solve this problem, we apply the variable neighborhood programming algorithm to the leaning datasets (see Algorithm 1). Each dataset concerns one district.

Algorithm 1: BasicVNP(T, k_{max})

repeat
  $k \leftarrow 1$
  while $k \leq k_{max}$ **do**
    $T' \leftarrow$ `Shake`(T, k)
    $T'' \leftarrow$ `LocalSearch`(T')
    `NeighborhoodChange`(T, T'', k)
  end while
until the stopping condition is met

In our work, we use the basic VNP variant [4, 5]. The solution is a program (function) that is presented as a tree. This tree graph includes two types of nodes: functional and terminal nodes, where the latter one contains constants and variables. The used functional set F includes arithmetic operators, which is $F = (+, -, *, /)$. The terminal set includes the problem's variables $\{x_1, x_2, x_3\}$ and constants that we decide to belong to interval $[-5,5]$.

The first solution is generated using the initialization method.

The following three neighborhood structures are used in our VNP method:

- $N(T)$ denotes neighborhood structure used in the local search. Each possible edge is added to the tree T that represents the current regression formula, and then transformation is performed to remove an edge from the circle that is obtained, in order to recover feasibility. Not all edges are allowed to be added or removed, as in usual tree with one node type (see [5] for details);

- $\mathcal{N}_1(T)$ or *Changing a node value*. It conserves the shape of the tree and changes only one value of a functional or a terminal node [4].
- $\mathcal{N}_2(T)$ *Swap operator*. In this neighborhood, a node has to be selected first from the current tree. Then, a new subtree is generated according to a chosen size. Finally, this new subtree is attached in the place of the subtree, corresponding to the selected node. More details are given in [4, 5].

Like the basic VNS, the basic VNP algorithm iterates *Shaking*, *Local search*, and *Neighborhood change* steps:

- **Shaking.** We use *Changing a node value* operator $\mathcal{N}_1(T)$ and the *Swap* operator $\mathcal{N}_2(T)$ that are randomly chosen for each k. Thus, the tree T' in kth neighborhood of T is obtained by repeating k times random move using either $\mathcal{N}_1$ or $\mathcal{N}_2$.
- **Local search.** We use the adapted elementary tree transformation, i.e., $\mathcal{N}_1(T)$ neighborhood is explored.
- **Neighborhood change.** If T'' represents better regression formula (with smaller error), then neighborhood parameter k is set to 1; otherwise, $k \leftarrow k + 1$.

4.4 Case Study on Life Expectancy at Russian Districts

In the past decade, Russia has been pursuing an active state policy in the field of health protection, increasing access to medical care. The average life expectancy at birth in the period from 2005 to 2018 increased by 7.5 years: from 65.37 to 72.91 [8]. However, the gap with the EU countries in this indicator in 2017 was 8.2 years (the European Union represented by 28 countries had 80.9 years [9], while Russian Federation had 72.7 years [8]).

There are various opinions of Russian experts on the impact of the health system on the life expectancy of the population, that can be found in [7]:

- Guzel Ulumbekova, the head of the Higher School of Organization and Management of Health Care, believes that with the quality and availability of medical care in Russia only 30% determine the life expectancy of the population, by 37%, it depends on socio-economic factors, primarily on income, and by 33% on lifestyle, in particular, on alcohol and tobacco consumption;
- Sergei Shishkin, the director of the Health Policy Center at the Higher School of Economics, also believes that life expectancy is determined by socio-economic conditions, lifestyle, and quality of medical care. However, hw concluded that there are no reliable methods to determine their proportion;
- Vitaly Omelyanovsky, the head of the Laboratory for the Evaluation of Health Care Technologies of the Russian Academy of People's State and the State Service under the President of the Russian Federation, believes that the world does not have a clear understanding of the quantitative contribution of medicine to the increase in life expectancy

4.4.1 One-Attribute Analysis

To carry out the study on the basis of the official data of Rosstat, a database for 2001–2016 has been compiled for 85 regions [14–19]. We choose from statistical tables just three economic indicators as mentioned before as three variables for our symbolic regression model (x_1, x_2, x_3) and life expectancy y. Here we first analyze each of those indicators separately. Then we analyze the results of our symbolic regression approach.

Life expectancy y at birth in Russia from 2001 to 2016 increased by 6.58 years from 65.29 to 71.87 years (Fig. 4.2, where each color represents different district). The minimum value increased from 56.48 to 64.21 years, and the maximum from 74.6 to 80.82 years. The gap between the minimum and maximum values during this period decreased from 1.3 to 1.26 times.

The major part of Russian citizens receive medical care for free [20] (according to the Russian Monitoring of the Economic Situation and Public Health of the Higher School of Economics in 2015).

Expenditure of territorial funds x_2 of compulsory medical insurance of constituent entities of the Russian Federation, at current prices, on average in Russia, increased from 0.6 to 11.1 million roubles per 1 thousand people (Fig. 4.3). The minimum value has grown from 0.005 to 7.5 million roubles per 1 thousand people,

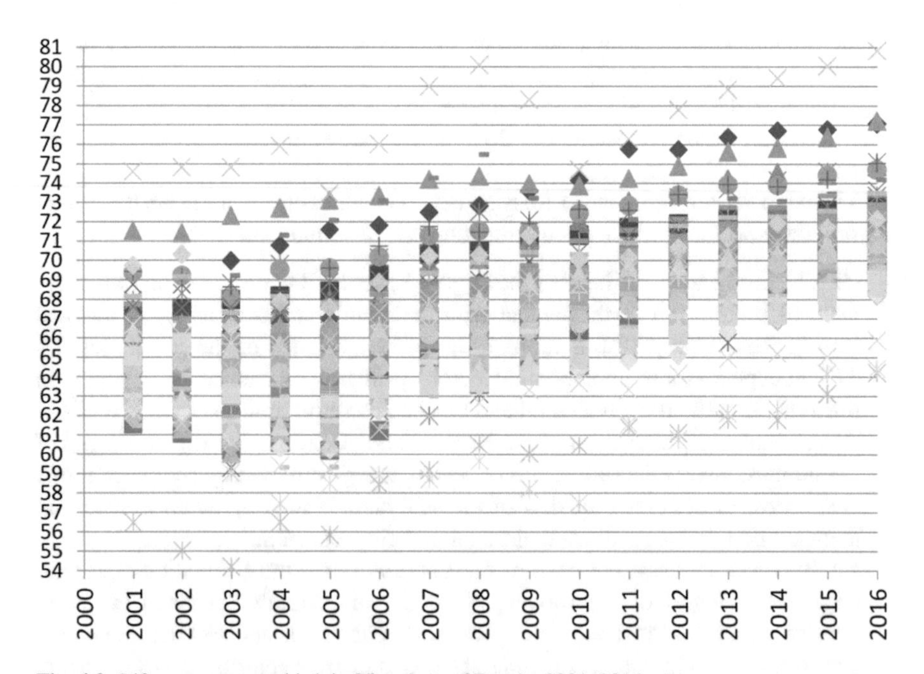

Fig. 4.2 Life expectancy at birth in 85 regions of Russia, 2001–2016

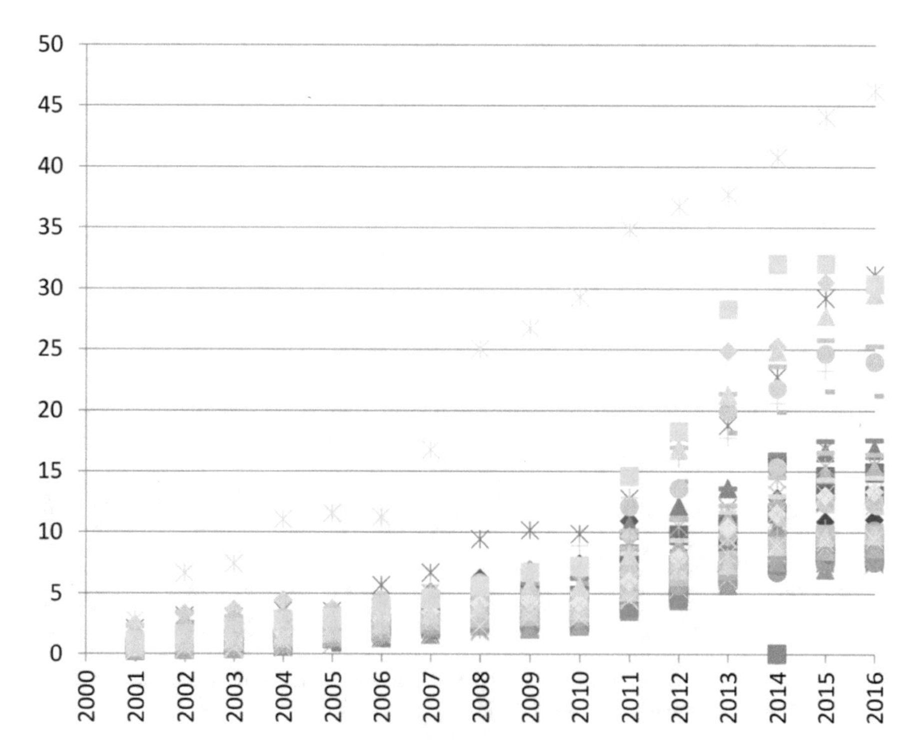

Fig. 4.3 Expenditure of territorial funds of compulsory medical insurance of constituent entities of the Russian Federation, million roubles per 1 thousand people (at current prices), 2001–2016

and the maximum from 2.8 to 46.3 million roubles per 1 thousand people. However, it is necessary to bear in mind that the data are given in current prices (Fig. 4.4).

Average per capita money income x_3 of the population in current prices on average in Russia increased from 3062 roubles per month to 30,744 roubles per month (Fig. 4.5). The minimum value increased from 909 to 14,107 roubles per month, and the maximum from 10,733 to 69,956 roubles per month. It must be borne in mind that the data are given in current prices.

4.4.2 Results and Discussion on Three-Attribute Data

We run VNP code to find formulas for each of the 85 districts. We start our analysis of the results obtained by VNP on Central Federal District.

Since we cannot present figure in four dimensions, we separate obtained life expectancy into three figures: in Fig. 4.6 y is plotted as a function of x_1 and x_2; then in Fig. 4.7 we present $y = f(x_1, x_3)$, and finally, in Fig. 4.8 we show dependency of life expectancy as a function of x_2 and x_3.

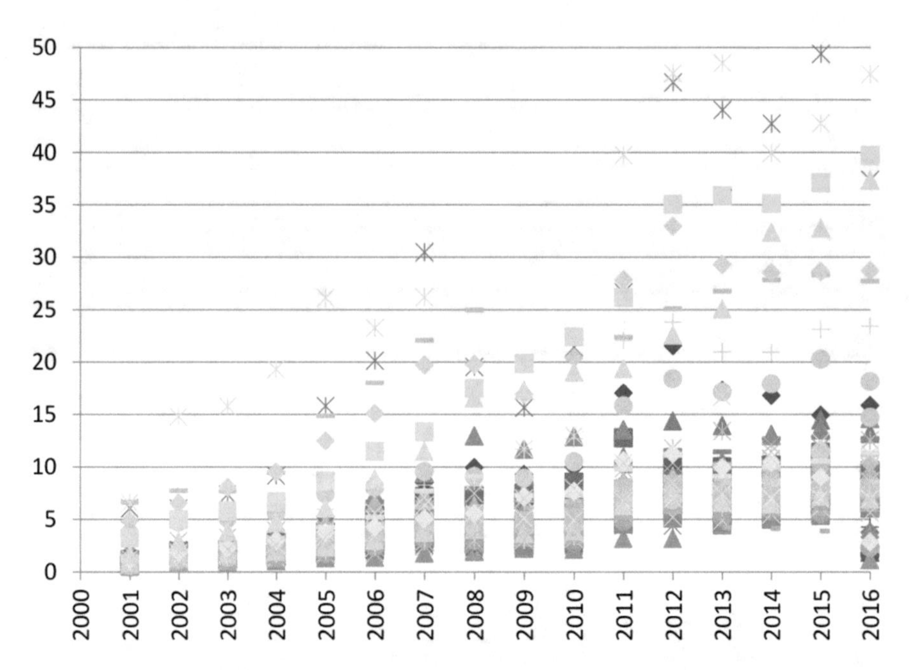

Fig. 4.4 Expenditures of the consolidated budgets of constituent entities of the Russian Federation on health, million roubles per 1 thousand people (at current prices), 2001–2016

It is clear that the final function $y = f(x_1, x_2, x_3)$ is almost linear. This observation is in accordance with other studies mentioned in Sect. 4.1.

The similar results are obtained by other 84 geographical districts. It is interesting to note that the output from our VNP code is sometimes very long and could be presented in reduced form, after elementary algebraic transformations. For example, for Ural Federal District obtained VNP output function is

$$f(x) = \frac{984067x_2}{50000\left(x_2 + \frac{7249}{1300}\right)\left(-\frac{62500x_1}{578641x_2} + \frac{281}{100} + \frac{95677}{10000\left(-\frac{22883}{2000} - \frac{8074}{625\left(-\frac{98863}{15625} + \frac{3003}{1060x_2}\right)}\right)}\right)} + \frac{6236451}{100000}$$

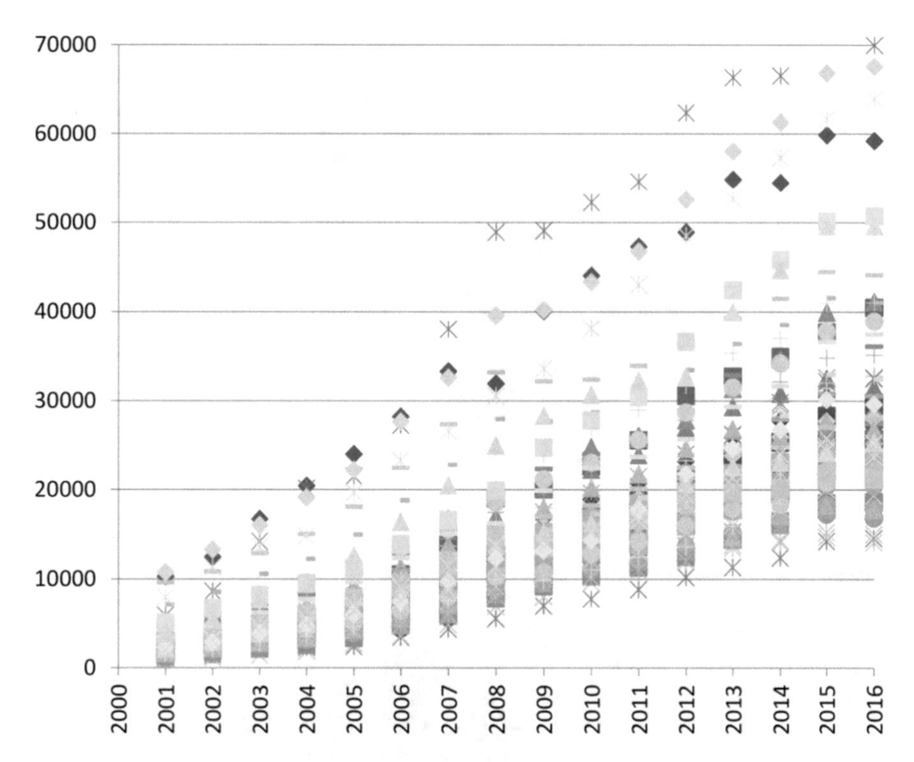

Fig. 4.5 Average per capita money income of population, per month, in roubles (at current prices), 2001–2016

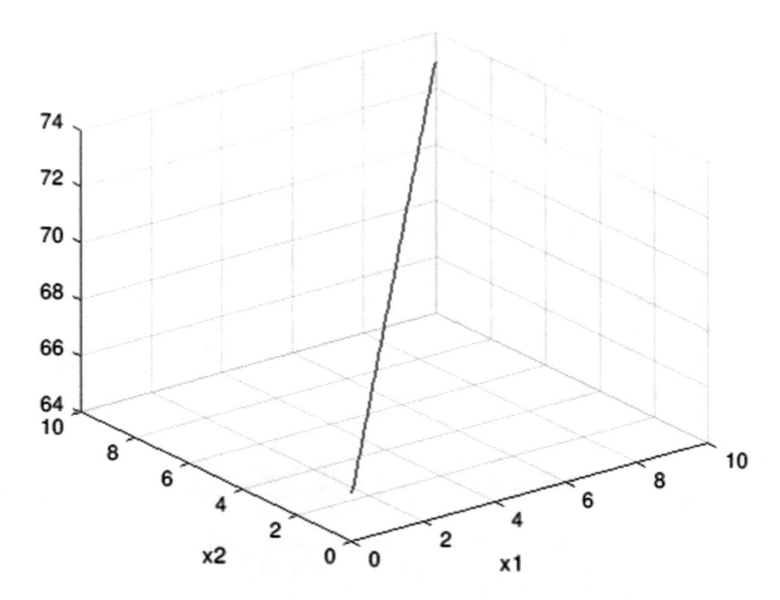

Fig. 4.6 Central Federal District output according to variables x_1 and x_2

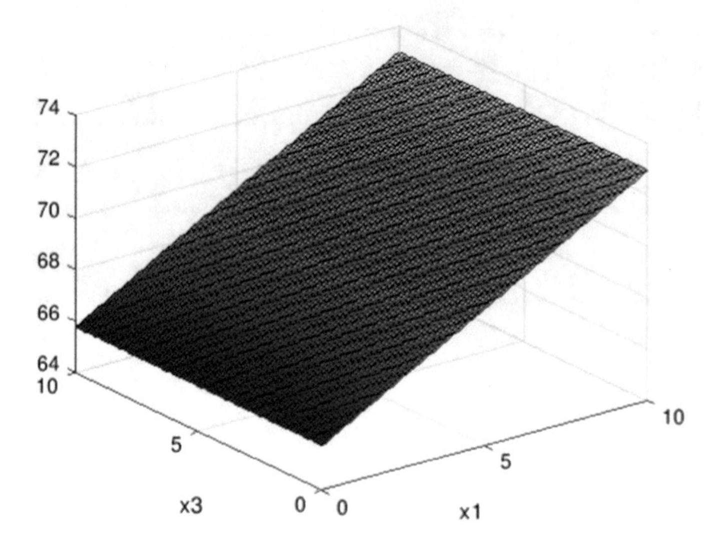

Fig. 4.7 Central Federal District output according to variables x_1 and x_3

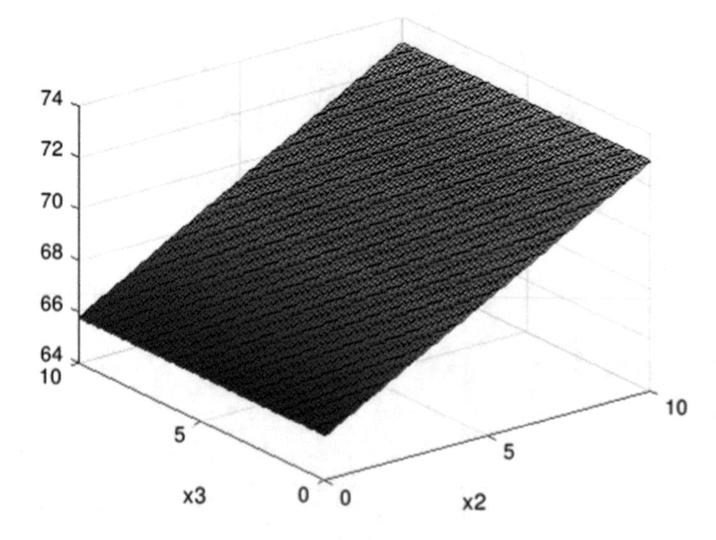

Fig. 4.8 Central Federal District output according to variables x_2 and x_3

that can be transformed to

$$
\begin{aligned}
f(x) &= 62.36 + \frac{19.68x_2}{(x_2+5.58)\left(2.81 - 0.11\frac{x_1}{x_2} + \frac{270.76-604.67x_2}{59.39x_2-32.37}\right)} \\
&= 62.36 \\
&\quad - \frac{1167.02x_2^3 - 637.04x_2^2}{438.04x_2^3 + 2264.46x_2^2 + 6.52x_1x_2^2 + 32.82x_1x_2 - 1003.28x_2 - 19.86x_1}.
\end{aligned}
$$

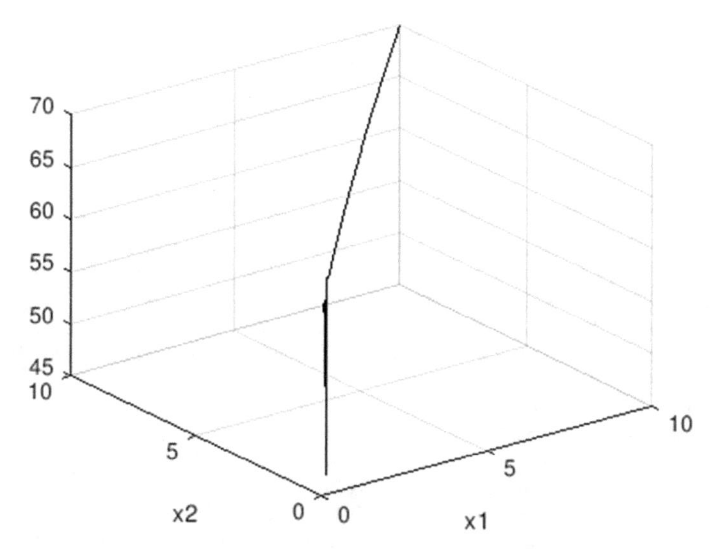

Fig. 4.9 Ural Federal District output according to variables x_1 and x_2

Even the function is not linear, after plotting it in (x_1, x_2, y) space (see Fig. 4.9) we see that from small x_1 and x_2 the almost linear function has one slope, and another for larger investment. Note also that x_3 is not included again in the formula.

4.5 Conclusions

In this chapter we addressed the life expectancy question and its relation with three expenditure types into a health-care system. We proposed symbolic regression and machine learning approach, to find whether there is some simple analytic dependence of life expectancy from the three expenditure types. We tested our approach on 85 Russian geographical districts, taking into account the data for 15 years. Our basic observations are:

(1) for small amount of investment, there are no clear functional dependencies of life expectancy and three investment types;
(2) for larger investment, there are almost linear dependencies in all 85 districts, i.e., the more the expenditure, the higher the expected life expectancy; and
(3) in some districts, one and sometimes even two types of investments have no influence on life expectancy. Most often variable x_3 did not have any influence on life expectancy.

Acknowledgments This publication is based on the work supported by the Khalifa University of Science and Technology under Award No. RC2 DSO. The research is also partially covered by the framework of the grant number BR05236839 "Development of Information Technologies and Systems for Stimulation of Personality's Sustainable Development as One of the Bases of Development of Digital Kazakhstan."

References

1. https://ourworldindata.org/grapher/life-expectancy-vs-health-expenditure
2. Lubitz J, Cai L, Kramarow E and Lentzner H. Life Expectancy, and Health Care Spending among the Elderly, *New England J of Medicine* 349 (2003) 1048–1055.
3. Jaba E, Balan C B, Robu I B. The Relationship between Life Expectancy at Birth and Health Expenditures Estimated by a Cross-country and Time-series Analysis. *Procedia Economics and Finance* 15 (2014) 108–114.
4. Elleuch S, Jarboui B and Mladenovic N, Variable neighborhood programming - A new automatic programming method in artificial intelligence, *G-2016-92*, 2016.
5. Elleuch S, Hansen P, Jarboui B and Mladenovic N. New VNP for Automatic Programming, *Electronic Notes in Discrete Mathematics* 58 (2017).
6. Lalonde, M. A new perspective on the health of Canadians. *Ottawa, ON: Minister of Supply and Services Canada. Retrieved from Public Health Agency of Canada* (1974). website: http://www.phac-aspc.gc.ca/ph-sp/pdf/perspect-eng.pdf
7. Life expectancy increasing in Russia, experts claim. https://www.vedomosti.ru/economics/articles/2018/05/29/770996-rosta-prodolzhitelnosti-zhizni
8. Russian statistical yearbook. *Rosstat* (2018) 84
9. Eurostat https://ec.europa.eu/eurostat/data/database
10. E. Shcherbakova. Life expectancy and health care in OECD countries. *Demoscope weekly* 757–758 (2018)
11. Health at a Glance 2017: OECD indicators, http://dx.doi.org/10.1787/888933602215
12. Health at a Glance 2017: OECD indicators, http://dx.doi.org/10.1787/888933602272
13. World health statistics 2017: monitoring health for the SDGs, Sustainable Development Goals *World Health Organization* (2017) 86–101.
14. Healthcare in Russia. 2017: Stat. book./Rosstat. - M., 2017. - 170 p.
15. Healthcare in Russia. 2015: Stat. book./Rosstat. - M., 2015. - 174 p.
16. Healthcare in Russia. 2011: Stat. book./Rosstat. - M., 2011. - 326 p.
17. Healthcare in Russia. 2009: Stat. book./Rosstat. - M., 2009. - 365 p.
18. Healthcare in Russia. 2007: Stat. book./Rosstat. - M., 2007. - 355 p.
19. Healthcare in Russia. 2005: Stat. book./Rosstat. - M., 2006. - 390 p.
20. Health care: current status and possible development scenarios: dokl. to the 18th Apr Int. scientific conf. on the problems of economic and social development, Moscow, April 11–14. 2017 - em M: Publ. House of the Higher School of Economics (2017) 8.

Chapter 5
Operating Theater Management System: Block-Scheduling

Bilal Bou Saleh, Ghazi Bou Saleh, and Oussama Barakat

Abstract In this article, we review the characteristics of the mixed integer linear programming (MILP) procedure to establish a block-schedule of an operating room (OR). We then propose another approach based on distributed artificial intelligence (DAI) to establish an optimal surgical program. This consists in optimizing the skeleton of a fixed block-schedule. Our main goal is to propose a flexible planning tool using the techniques (DAI) to create or improve periodically a block-schedule. Another motivation of our study is to design a model that allows planning that can be adjusted to changes in the state of the operating theater. In fact, the OR schedule established by conventional methods with non-real data does not take into account the weekly variability, however common, especially the variations of the doctors' preferences concerning their availability for a time slot of the week, the amount of hours awarded to a team of surgeons or a given specialty, the duration of shifts, and the availability of a specific operating room for a specific team. Simulation tests on a typical case using real data are performed by both methods. The results allow us to conclude as to the superiority of the model (DAI).

B. B. Saleh (✉)
University of Bourgogne Franche Comté, Besançon, France

Lebanese University, Beirut, Lebanon

G. B. Saleh
Lebanese University, Faculty of Technology- SAIDA, Saida, Lebanon

O. Barakat
Nanomedicine LAB, University of Bourgogne Franche-Comté, Besançon, France

Faculté of Sciences and Technologies, Besançon, France
e-mail: oussama.barakat@univ-fcomte.fr

M. Masmoudi et al. (eds.), *Artificial Intelligence and Data Mining in Healthcare*,
https://doi.org/10.1007/978-3-030-45240-7_5

5.1 General Context

5.1.1 Introduction

Modern hospitals must provide their patients with efficient medical services while seeking high profitability by optimizing the use of their resources. These goals are ubiquitous for the operating room administrator, who has the very influential task of sharing the hours of operation of the OR between his surgeons. In fact, hospital planning procedures often pose logistical challenges because of a wide variation of parameters, for example: the availability of surgeons, their preferences, and the number of rooms, their equipment, and the opening ranges of the operating theater. With a very dynamic RO state and a multitude of factors to consider, planning surgical procedures is not an easy task. Optimizing the surgical schedule dramatically improves the efficiency of the hospital's surgical service by reducing costs, increasing operating room utilization and patient convenience by reducing wait times. We will conduct our study for a renowned "University Hospital Center" (CHU) in Lebanon. The hospital excels in the services of cardiology, neurology, oncology, general surgery, and obstetric surgery, with an ambulance service based on the hospital for emergency transport. As is common practice in this area, CHU uses a "block-scheduling" system to program surgical procedures. It is a goal-based programming system [1] that assigns a block of time and place to each team of doctors for a given period of time (usually one week). Each block is reserved for the exclusive use of the specified owner and only if the time slot is not used, it is made available to other doctors.

5.1.2 CHU Operating Theater "Block-Schedule"

At the beginning of this study, we obtained the used version of the block-schedule in the operating theater of the CHU. As things stand, this schedule with lack of precision does not plan all surgical teams, but keep holes in the calendar to be filled manually at the end of each week for the next week. In particular, the calendar includes many unused slots, with an overall utilization rate of about 60%. Part of our goal is to increase overall utilization; this involves planning in advance the maximum possible number of surgical teams while trying to stay not far away to the time required by each team. The format must also be modified to improve the readability of the block-schedule of CHU operating theater.

5.1.3 Search Background

Optimizing the operating theater schedule is a fairly classic problem in hospital settings. In the past, many studies using the principles and means of operational research have been conducted. Depending on the objectives we aim for, different formulations of the objective function to be optimized have been proposed. In [2], the authors use a linear programming technique to solve the assignment of operating rooms by maximizing the function representing the financial performance of the operating theater. The authors take into account a series of parameters such as the operating hours of the operating room, the duration of the surgical procedures to be programmed, and the operating costs of the operating room (equipment costs and fees of the HR). Excel Solver is used as a means of calculation to solve the optimization problem. Blake et al. [3] suggest that operating theater managers use a tool based on the integer programming method to assign blocks of time to surgical teams. The fitness function to minimize is the difference between the real time allocation of the operating theater and the target allocations desired by each team (predetermined by the teams or following a calculation of contribution to the performance). In [4], the authors use a mixed linear programming MILP method to assign capacity in the operating room, with optimization focused on reducing the length of stay of patients. Reducing the total patient's hospitalization time is in the interests of the hospital, because costs in relation to hospital's staying are the most important expenses. This leads to the optimization of the use of available space, in other words the occupancy rate. The data that we obtain from the CHU hospital did not contain precise cost information; we decide to assign an arbitrary hourly rate to normal working hours. We then applied a linear increase in costs for overtime interventions. We focused on the use of operating rooms. We use historical data to establish probability functions for target of weekly allocations for teams of surgery and a cooperative multi-agent planner system to optimize the planning skeleton containing the blocks assigned each week to each team.

5.1.4 Problem Definition

Determining an equitable time allocation method among surgeon groups is a common problem in hospitals with limited time capacity in the operating room. Operating theater managers begin by determining how much time should be allocated to each surgical team, based on their experience and expertise, or by using the formulas based on overall performance optimization from the partial margins contribution of these teams. Once the time shares of each group have been calculated, a method is needed to adapt an existing "OR" schedule or to generate a new one. In this study, we closely examine the methods of allocating slots, during the days of the week, to specific surgical groups. We assume that the number of hours of operation "OR" to assign to each team is a predetermined datum of the problem to be solved.

5.2 MILP Problem Formulation

To create a trial version of the problem and present the corresponding formulation, we followed the steps prior to establishing the objective function and introduced a set of restrictions to make the problem soluble. We used goal assignments for each surgical group as pre-established data and made some assumptions about the types of operating rooms.

5.2.1 *Definition of Decision Variables*

For this trial, there are eight types of operating rooms in this hospital, each based on an example of a particular weekly schedule. There are 5 days of operation per week, as well as nine surgical teams that will be assigned to the main program. This gives us a total of $12 \times 12 \times 9 = 1296$ variables, $x_i jk$, where

$$xijk = OR\, number\ \ type\,(i) \in \{1,\ .., ntype\}$$
$$assigned\ to\ the\ surgical\ team\ (j\ ) \in \{1,\ .., nteam\}$$
$$on\ the\ shift(k) \in \{1,\ .., nshift\}$$

"i" indicates the type of operating room that will differentiate those that are exclusively specialized in performing certain types of surgical procedures. "j" indicates the surgical group (see test data). The index k indicates the reference of a working time shift. There are 12 time shifts: one for AM and one for PM from Monday to Saturday (k ranging from 1 to 12).

5.2.2 *Objective Function*

The chosen optimization consists of minimizing the reduced and normed difference between the targets hours requested and the hours effectively allocated and that for all the surgical teams and for the planned week. We have included in the objective function a "formulation term" to introduce penalty weights for Saturday and Friday teams, in order to limit our model of allocating unappreciated shift-hours to weekend and Friday teams. This amounts to minimizing the following function:

$$F = \Delta + \alpha\, \partial \tag{5.1}$$

$$\Delta = \sum_{j=1}^{j=nteam} \frac{\max(0,\ t_j - \sum_{i=1}^{ntype} \sum_{k=1}^{nshift} d_{ik} x_{ijk})}{t_j} \tag{5.2}$$

$$\partial = \sum_{j=1}^{j=nteam} \frac{\sum_{i=1}^{ntype} \sum_{k=9}^{12} d_{ik} x_{ijk})}{t_j} \tag{5.3}$$

with t_j as the total number of hours requested by surgical team "j," $d_i k$ as the hours amount of operation staffed in OR type i during shift k. The number of hours required by the different surgical groups being very variable, the choice of the objective function ensures some equity between the surgical teams by minimizing the reduced normed difference rather than the absolute difference.

5.2.3 *Constraints*

1. At any work-shift k, and for any OR type, the total number of planned operating rooms must not exceed the total number of available rooms of this type.

$$\forall\, k,\ \forall\, i,\ \sum_j x_{ijk} \le R_i$$

2. Sum of the ORs of all types allocated to team j at each work-shift k must be less than the total number of doctors in that team on that work-shift (S_{jk}).

$$\forall\, j,\ \forall\, k,\ \sum_i x_{ijk} \le S_{jk}$$

For our test example we consider an operating theater with twelve surgery rooms and eight types of operating room. In these operating rooms, nine teams operate and are trained with Sj surgeons.
To take into account the possible interruption times between the interventions, the number of hours effective in each work-shift has been retouched. Thus the AM work-shift is reduced from 5 h to 4.5 h and the PM work-shift is reduced from 3 h to 2.25 effective hours and this for each working day from Monday to Saturday. This gives a weekly capacity of 12 operating rooms totaling 486 available operating hours (compared to the target weekly demand of 566 h).

5.2.4 *Results of the Simulation*

Using the presented formulation and performing a simulation with the data selected for this test example, the obtained schedule meets our expectations, there is no overlap of shifts, and the two specialized ORs provide the team of surgery that they are qualified to serve. Moreover, we obtained the following results: the total, for all surgical groups of the weekly hours allocated, is 479.25 h for a total of 566 h requested. The utilization rate (effective work time allocated/maximum capacity in hours) is of 78.3%. We have developed a timetable format, with the days of the week in lines, the ORs (with their type of color code) in columns, and work-shifts of interventions associated with each team (in color code) in the body of the timetable.

MILP		OR1	OR2	OR3	OR4	OR5	OR6	OR7	OR8	OR9	OR10	OR11	OR12
MON	AM	E	E	E	E	E	E	E	B	B	C	F	G
	PM	A	E	E	E	A	H	A	B	G	C	F	A
TUES	AM	A	A	E	E	E	E	E	I	H	C	F	H
	PM	E	I	H	H	B	G	D	D	D	C	F	D
WED	AM	E	E	E	E	F	F	F	F	F	C	F	A
	PM	E	E	F	F	F	F	I	H	B	C	F	B
THURS	AM	A	E	F	F	F	F	F	A	A	C	F	B
	PM	F	F	F	F	F	F	I	H	B	C	F	B
FRI	AM	E	E	E	E	E	E	E	E	E	C	F	E
	PM	A	E	E	E	E	E	I	B	B	C	F	G
SAT	AM	E	E	E	E	H	B	B	D	D		F	D
	PM	A	E	E	E	E	A	A	B	A		F	D

MILP	Specialty	Target operating-time	Planned operating-time	Ratio
Team E	General Surgery	TE = 191 hours	TE = 173.25 hours	90.7 %
Team H	Otolaryngology	TN = 30 hours	TN = 24.75 hours	82.5 %
Team C	Ophthalmology	TU = 40 hours	TU = 33.75 hours	84.3 %
Team D	Gynecology	TW = 30 hours	TW = 24.75 hours	82.5 %
Team G	Urology	TS = 18 hours	TS = 11.25 hours	62.5 %
Team B	Pediatric Surgery	TP = 62 hours	TP = 42.75 hours	68.9 %
Team F	vascular, cardiac, thorax	TF = 121 hours	TF = 108 hours	89.2 %
Team A	Digestive	TA = 60 hours	TA = 47.25 hours	78.75 %
Team I	Neurosurgery	TI = 14 hours	TI = 13.5 hours	96.4%
All		T = 566 hours	T = 479.25 hours	84.6 %

5.3 MAS Planner Approach

5.3.1 Preface

Our motivation is to design a model that allows planning that can be adjusted to changes in the state of the operating theater. Indeed, a static surgical program does not take into account the variability of the weekly program, in particular the variation of the surgeon's preferences for the choice of the day and the intervention room, the number of hours allocated to a team of surgeons, the duration of work-shifts, the number of surgeons available in parallel with each shift, and the requirement that a type of surgery be performed in a particular operating room. The planning approach that uses the principles of distributed artificial intelligence is a different and alternative approach to those usually used. This method has considerable potential for solving management problems. In what follows, we propose a multi-agent intelligent system architecture (MAS) that noticeably models the operating room by doing so that each element involved in the process is represented by a specific agent. The so-called multi-agent system is conventionally constituted of a society of several intelligent agents that are autonomous entities, human or software, communicating and proactive [5]. MAS has recently been proposed as an appropriate management approach for areas such as multi-robot systems [6], electronic commerce [7], security, manufacturing [8], etc.

5.3.2 Multi-Agent Planner

A new application of artificial intelligence emerges, the multi-agent planners' methodology which involves integrating planning capabilities into the society of intelligent agents. Thus the multi-agent system will be able, through communication and cooperation between its agents, to develop an action plan established to achieve the common objectives. The advent of MAP will involve a considerable expansion of the community of automatic planning methods [9]. MAP can give two types of plans: the global program which is common to a company of several agents or the specific program for each agent but implemented jointly by all the agents. A fully cooperative MAP approach, similar to the one adopted, states that all agents are fully cooperative in the meaning that only the overall objectives of the joint venture count. In our multi-agent planners' we have restricted the decision-making ability to the sole manager-agent. The mechanism of distribution of the reservations of the time slots is carried out by a so-called global heuristic. We present in the following, the architecture of the planner system used to solve the issue of distribution of time slots reservations on the 12 rooms of the operating theater. The management system consists of two main layers: the physical layer including surgeons or specialty services, the operating theater administrator and the surgery rooms; the distributed

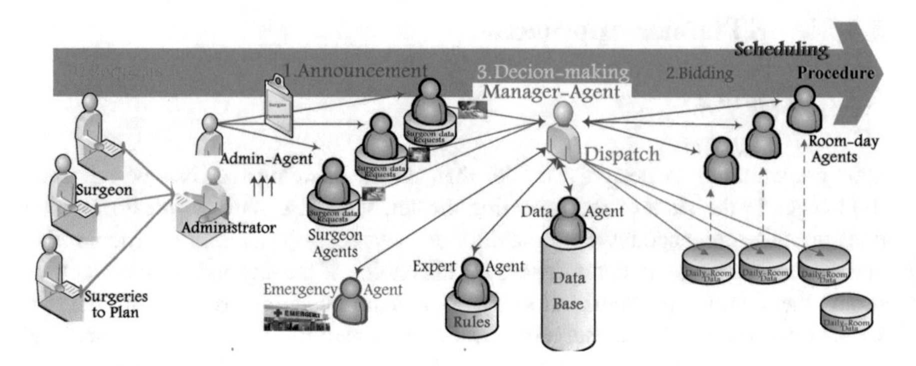

Fig. 5.1 Multi-agent planner system

artificial intelligence layer consists of a multi-agent system whose architecture we will shown in the Fig. 5.1.

MAS consists of:

- **Agent-Emergency:** embodies the emergency department, it has a specific graphic interface that is used to announce requests for emergency surgeries.
- **Agent-Admin** represents the administrator of the operating theater, responsible for entering into the system all requests for time slots reservations for surgeries or, on the contrary, making cancellations.
- **Agent-Manager** is the decision-maker of the system, each of these decisions, to assign a time slot reservation to a surgical team, aims to optimize the overall performance sought. Predefined criteria and rules are the basis of his decisions. It is the only initiator of the two ST and CNP communication protocols used in MAP to establish optimized schedule. For each request to insert a reservation for a surgical intervention in the planning, agent-manager enriches the request with additional data relating to the constraints to be respected. Then, it will launch a CNP call for tenders for possible assignments. It decides to allocate this reservation in the best slots available.
- **Agent-Surgeon** embodies a team of surgeons, a single surgeon, or even a surgery department. Each surgeon-agent communicates with agent-admin and retrieves the target list, coming from the team he represents, of the surgeries to plan. All the surgeon-agents communicate and cooperate with the agent-manager, according to an iterative process, to find the best locations in the plan of the time slots corresponding to their target list.
- **Agent-Room:** Each agent-room embodies a physical surgery room. Each RA is responsible for collaborating with agent-manager to establish their own schedule. Thus, the ARs' participate in calls for tender of the CNP protocol initiated by AM. They use their local planning software to supplement or adjust their schedule. The decision to reserve a time slot for a given team is the sole responsibility of the agent-manager. Once the MA has decided the AR must then update its schedule being filled.

- **Agent-Data:** has the task of filtering, classifying, and storing data in a predefined format. It is obviously responsible to extract this data if necessary.
- **Agent-Expert** accumulates the data in relation to know-how, history, and experience. It records the laws and rules relating to the jurisdiction of the operating theater. It provides these data when needed.

5.3.3 *Patient's Programming*

This article presents a planning approach corresponding to the patient's programming. Patient programming is an operating room management module, located upstream of the surgical intervention process [10]. It consists in reserving, according to the surgeons' requests, the possible time slots for the interventions that are planned for a period conventionally of a week. In [11] the authors propose simply a waiting list generation approach. Otherwise, according to Marcon there are three models of patient programming [12] :

- **Open Booking:** the programming is centralized at the level of the manager of the operating room. Reservations are made in chronological order, as the requests arrive or periodically. Organizing this type of planning is simple. Nevertheless, this can lead to malfunctions of under-utilization of resources or, on the contrary, to many overtime hours.
- **Block-scheduling** consists of developing, from all the requests of the surgical teams, a skeleton program. This initial schedule consists of bookings of well-defined time slots and each associated with a surgical team. The surgical teams then use these reserved time slots to plan their elective interventions.
- **Modified block-scheduling:** similar in principle to block-scheduling, but using time slots of adjustable duration, with the possibility of extension if necessary.

5.3.4 *Frequency Evaluation*

To automatically develop certain indicators relating to experience and know-how, the agent-expert records the data and parameters of each surgery after completion. These data can be used to assess the frequency of each type of surgery to be performed in a given time slot. In this work, two of these indicators are introduced:

1. Normed absolute ability: by doing a statistics on several months ago, let Ns (j, ks) be the occupations number relative to team j realized in time slot ks, let #(j, ks) be the normed absolute ability that is calculated as follows:

$$\#\,(\mathrm{j},\ \mathrm{ks}) = \frac{\mathrm{Ns}(\mathrm{j}, \mathrm{ks})}{\sum_{\mathrm{j}=1}^{nteam} \mathrm{Ns}(\mathrm{j}, \mathrm{ks})}$$

2. Relative ability: Let \$(j, ks) be the relative ability of the time slot ks to receive a reservation relative to team j that is calculated as follows:

$$\$ (\mathrm{j},\ \mathrm{ks}) = \frac{(\mathrm{j},\mathrm{ks}) - \overline{(\mathrm{j},*)}}{\max\ (\mathrm{j},*) - \min\ (\mathrm{j},*)}$$

Time slots with a relative ability positive are those with a higher than average ability to receive a reservation relative to team j.

5.3.5 *Surgeon's Preferences*

Surgeon teams have preferences regarding the time slots to be reserved for their interventions. It is obvious that if surgeons impose all the slots of their choice, planning is not possible because there will certainly be a lot of collisions in the schedule. MAP "Block-Scheduling" is designed for a context in which surgeons are totally cooperative in order to achieve the common goal of improving the overall efficiency of the operating theater. This means that all teams accept the optimized outcome of the planning tool, which may, for some slot reservations, not match their maximum preference. Each team is associated, in addition to its target time request, with a preference matrix Pref (team, time slot). This matrix indicates in Pref (j, ks) a metric corresponding to the weight of the preference of the team j request to reserve the time slot ks. This matrix can be done by teams, once and for all; it will be stocked in the database and will be changed only in case of change of team's preference. Nevertheless, this functionality requires for a real application case that the teams are more attentive in the definition of their preferences. In other words, the preference matrix must not contain only (1) and (0) but well-calculated weights. This gives flexibility and allows the planning process to soften further and gives it the degree of freedom needed to find the optimal solution.
In addition, a surgery team may have higher priority than others for occupying a room. To be able to take into account this distinction, we have defined a matrix containing in Pri (j, n) the priority, given by the administrator, for the team j to use the room n. We admit here that several teams may have the same priority to use a given room.

5.3.6 *Virtual Cost*

Optimizing the room occupancy rate may require assigning a team a time slot that does not match their preferences. We believe that this type of assignment should be rare but possible. However, the dissatisfaction of the team of surgeons, as for the reservation of a time slot not corresponding to its maximum preference, must be taken into account in the calculation of the cost. To introduce the concept of virtual cost of a reservation of a time slot, consider the following example: Let A correspond to placing in the time slot (ks) a reservation of (x) hours for an intervention of the team (j) surgeons. Let (p) the weight of this allocation. Let rcost (A) be the actual cost of the (x) hours of operation. We define the virtual cost of A as follows:

$$\alpha \in [0,\ 1]$$

$$p = \alpha\,(\mathrm{j},\ \mathrm{ks}) + (1-\alpha)\mathrm{Pref}\,(\mathrm{j},\mathrm{ks})$$

$$vcost = rcost\ + (1 - p) * rcost$$

This amounts to adding a fictitious extra cost corresponding to the non-preference of the surgeons. Note that the block-scheduling software is designed with the convenience of surgeons in mind. This led us to use the virtual cost minimization criterion as a basis for decision.

5.3.7 Block-Scheduling Algorithm

In what follows we will explain the planning mechanism to create a new schedule. The methodology is characterized by the existence of a single "decision-maker" (agent-manager), who controls the mechanism and decides on the configuration of each time slot when filling the schedule being drawn up. Agent-manager receives a request to allocate a reservation from an agent-surgeon. Agent-manager enriches the request by the information, concerning the preferences, the constraints, and the type of room, obtained from agent-expert. He issues a call for tenders (CNP) to agents-room who can participate in this negotiation. Each agent-room calculates the virtual cost of each of these time slots still available and selects the best one to answer agent-manager by indicating the corresponding virtual cost. Agent-manager analyzes the responses received. He selects the one that has the smallest virtual cost and that respects the constraints regarding the number of surgeons who can be present in parallel. He informs the agents-room of his decision and asks the winner to include this reservation in his schedule. The planning is distributed in iterations on the requests of the surgeon-agents. At the end of a round to include a reservation, agent-manager receives a new request from the next agent-surgeon. He proceeds in the same way to find a location in the planning. Iterations for filling bookings end when the list of received requests is empty (all requests have been scheduled) and when unplanned requests cannot be fulfilled.

5.3.8 Optimization Algorithm

This algorithm processes an existing calendar to find acceptable time slots for inserting one or more reservations. When manager-agent fails to place a submitted reservation in a regular cycle, this algorithm is used to attempt to place a reservation, replacing another with less priority. The optimization algorithm is located at the agent-manager level. This one tries to place, iteratively, the unfitted reservations. For each unplanned request, the algorithm analyzes the contents of the slots in the existing schedule. It lists all slots that may be suitable and occupied by other reservations of lower priority. It selects the one with the highest virtual cost. He replaces it with the one he seeks to place. The dislodged request enters the list of unplanned requests. The algorithm stops when no replacement is possible.

5.3.9 Performance Metrics

Block-scheduling planner software is designed to optimize the occupancy rate of surgery while emphasizing the "convenience of the surgeon" in each decision. To measure the performance of the proposed method, we create indicators that are calculated automatically:

1. The satisfaction rate.

$$SI(j) = \frac{\sum_{ks=1}^{Nslot} \mathrm{S}\,(\mathrm{j, ks}) * \mathrm{Pref}\,(\mathrm{j, ks}) * \mathrm{d}\ \ (\mathrm{ks})}{t_j}$$

2. The performance rate.

$$PI(j) = \frac{\sum_{ks=1}^{Nslot} \mathrm{S}\,(\mathrm{j, ks}) * \mathrm{d}\ \ (\mathrm{ks})}{t_j}$$

3. The occupation rate.

$$OI(j) = \frac{\sum_{ks=1}^{Nslot} \mathrm{S}\,(\mathrm{j, ks}) * \mathrm{d}\ \ (\mathrm{ks})}{\sum_{ks=1}^{Nslot} \mathrm{d}\ \ (\mathrm{ks})}$$

With:

$$\mathrm{S}\,(\mathrm{j, ks}) = 1\, if\ j\ \ occupies\, ks,\ 0\, Elsewhere$$

$$d\,(ks)\ \ duration\ of\ ks$$

$$Nslot = 12 * 12\, total\ number\ of\ timeslots$$

5.4 Experimental Test

5.4.1 Test Data

After reviewing 6-month-old weekly schedules at CHU hospital in Lebanon, we define a typical problem to test the proposed planning approach. The hospital currently uses a fixed block-schedule. It has 12 operating rooms, 2 of which are experts in specific interventions and ten can handle all types of surgery. Officially, the opening hours of these rooms are all week between 8:00 and 16:30 except for non-working Sundays. Each workday is divided into 2 teams, AM and PM. To take, more or less, into account the hazards of the operating room process, such as emergency surgeries, each AM shift is reduced to 4.5 h and each PM shift to 2.25 h, for a total of 6.75 h per day of work. Otherwise, the data reveals about fifty surgeons who intervene in these operating theaters and are affiliated to 9 separate surgical teams. Ten ORs (type 1) have the capability to provide surgeons doing any type of surgery. The other 2 rooms can only serve specific surgical teams. More precisely, OR 10 (type 2) is intended for the surgical team C and OR 11 (type 3) is intended for the surgical team F. In addition, every surgeon who intervenes in an OR is associated with a particular team of surgery. This will indicate that he is specialized in performing certain types of surgical procedures. For greater flexibility, all surgeons in a surgical team are assumed to be interchangeable and for the example treated we have adopted the following classification:

	Specialty	Target operating-time	Rooms
Team E	General Surgery	TE = 191 hours	Type 1= all – OR (11, 10)
Team H	Otolaryngology	TN = 30 hours	Type 1 + Pref(R7)
Team C	Ophthalmology	TU = 40 hours	Type 2 = OR 10
Team D	Gynecology	TW = 30 hours	Type 1 + Pref(OR9)
Team G	Urology	TS = 18 hours	Type 1 + Pref(OR2)
Team B	Pediatric Surgery	TP = 62 hours	Type 1 + Pref(R8, R12)
Team F	vascular, cardiac, thorax	TF = 121 hours	Type 3 =R11 + Pref(R1, R3)
Team A	Digestive	TA = 60 hours	Type 1
Team I	Neurosurgery	TI = 14 hours	Type 1 + Pref(OR8)

The availability of surgeons for particular periods must be considered in the planning model. Some surgeons may not want to work on Saturdays or Fridays, which reduces the number of doctors available at these times. Some surgeons are not available to work certain working days due to other professional obligations. Data about preferences of surgeons will be added to all constraints by adjusting the number of doctors who can work at the same time on each work-shift.

J / k	Team E	Team H	Team C	Team D	Team G	Team B	Team F	Team A	Team I
1	6	1	1	1	1	2	3	2	1
2	3	1	1	1	1	2	3	2	1
3	6	1	1	1	1	2	3	2	1
4	3	1	1	1	1	2	3	2	1
5	6	1	1	1	1	2	3	2	1
6	3	1	1	1	1	2	3	2	1
7	6	1	1	1	1	2	3	2	1
8	3	1	1	1	1	2	3	2	1
9	6	1	1	1	1	2	3	2	0
10	3	1	1	1	1	2	3	2	0
11	6	0	0	0	0	2	3	2	0
12	3	0	0	0	0	2	3	2	0

5.4.2 *Simulation Results*

Using the MAS planner and performing a simulation with the data presented for this test example, the schedule obtained corresponds perfectly to the optimal planning sought. There is no problem regarding the number of surgeons present in parallel and all rooms serve the surgical team according to their preference or the constraints already expressed. To present the schedule obtained in an easy to understand way, we drew a timetable format, with the days of the week in lines, the ORs (with their type of color code) in columns, and work-shifts of interventions associated with each team (in color code) in the body of the timetable.

MAS planner		OR10	OR11	OR1	OR3	OR4	OR5	OR6	OR2	OR7	OR8	OR12	OR9
				Pref					pref	pref	pref	pref	pref
MON	AM	C	F	F	E	E	E	E	G	H	B	E	D
	PM	C	F	F	F	E	E	A	G	H	B	A	D
TUES	AM	C	F	F	E	E	E	E	A	H	I	B	D
	PM	C	F	F	F	E	E	A	G	A	I	B	D
WED	AM	C	F	F	E	E	E	E	E	H	B	A	A
	PM	C	F	F	F	E	E	A	G	A	I	B	D
THURS	AM	C	F	F	F	E	E	E	E	H	B	B	A
	PM	C	F	F	E	E	A	A	G	H	I	B	D
FRI	AM	C	F	F	E	E	E	E	E	A	B	B	D
	PM	C	F	F	F	E	E	A	A	H	B	B	D
SAT	AM		F	F	F	E	E	E	E	E	B	B	E
	PM		F	F	F	E	E	E	E	E	B	A	A

The table below shows the performance indicator of the block-scheduling software obtained from the planning test performed. The performance rate averages 84.6%. In this test, the satisfaction rate is the same as the performance rate because all the reservations that have been planned correspond to the preferences of the surgeon. It should be noted here that the total demand is 566 h, while the total capacity is 486 h. It is obvious that block-scheduling offers the best possible solution and can only solve the problem of exceeding the capacity limit by refusing a set of requests.

MAS planner	Specialty	Target operating-time	Planned operating-time	Ratio
Team E	General Surgery	TE = 191 hours	TE = 164.25 hours	85.9 %
Team H	Otolaryngology	TN = 30 hours	TN = 24.75 hours	82.5 %
Team C	Ophthalmology	TU = 40 hours	TU = 33.75 hours	84.3 %
Team D	Gynecology	TW = 30 hours	TW = 24.75 hours	82.5 %
Team G	Urology	TS = 18 hours	TS = 13.5 hours	75 %
Team B	Pediatric Surgery	TP = 62 hours	TP = 56.25 hours	90.7 %
Team F	vascular, cardiac, thorax	TF = 121 hours	TF = 101.25 hours	83.6 %
Team A	Digestive	TA = 60 hours	TA = 49.5 hours	82.5 %
Team I	Neurosurgery	TI = 14 hours	TI = 11.25 hours	80.3 %
All		T = 566 hours	T = 479.25 hours	84.6 %

5.5 Conclusion

On the basis of the results obtained, we believe that the MILP and MAS models work well and can be used as tools for effective planning. Both models are configured to allow adjustment of certain parameters. Nevertheless, this is where the significant difference between the two models lies. The MILP model transforms the constraints into equations and seeks the optimal solution by solving the whole system of equations. The more constraints we add, the more complicated the system will be and it will be difficult to find the optimum. While in MAS, constraints become rules to respect. Their number does not affect the difficulty of finding the optimal solution. In addition, MAS finds the optimal overall solution by deciding, iteratively, on the best location for each "individual booking." So MAS processes one reservation at a time, which makes it much simpler to find the optimal solution. All this allows us to conclude that the MAS approach is much better for OR planning because it can easily accommodate a wide range of constraints.

References

1. I. Ozkarahan, "Allocation of Surgeries to Operating Rooms by Goal Programing", J. Med. Syst., vol. 24, no 6, p. 339–378, déc. 2000.
2. P. C. Kuo, R. A. Schroeder, S. Mahaffey, et R. R. Bollinger, "Optimization of operating room allocation using linear programming techniques", J. Am. Coll. Surg., vol. 197, no 6, p. 889–895, déc. 2003.
3. J. T. Blake, F. Dexter, et J. Donald, "Operating Room Managers' Use of Integer Programming for Assigning Block Time to Surgical Groups: A Case Study", Anesth. Analg., vol. 94, no 1, p. 143, janv. 2002.
4. B. Zhang, P. Murali, M. M. Dessouky, et D. Belson, "A mixed integer programming approach for allocating operating room capacity", J. Oper. Res. Soc., vol. 60, no 5, p. 663–673, mai 2009.
5. N. R. Jennings, P. Faratin, A. R. Lomuscio, S. Parsons, M. J. Wooldridge, et C. Sierra, "Automated Negotiation: Prospects, Methods and Challenges", Group Decis. Negot., vol. 10, no 2, p. 199–215, mars 2001.
6. J. Ota, "Multi-agent robot systems as distributed autonomous systems", Adv. Eng. Inform., vol. 20, no 1, p. 59–70, janv. 2006.
7. R. H. Guttman, A. Moukas, et P. Maes, "Agent-mediated electronic commerce: a survey", Knowl. Eng Rev., vol. 13, p. 147–159, 1998.
8. W. Shen, Éd., Information Technology for Balanced Manufacturing Systems: IFIP TC 5, WG 5.5 Seventh International Conference on Information Technology for Balanced Automation Systems in Manufacturing and Services, Niagara Falls, Ontario, Canada, September 4–6, 2006. Springer US, 2006.
9. N. T. Nguyen et R. P. Katarzyniak, "Actions and social interactions in multi-agent systems", Knowl. Inf. Syst., vol. 18, no 2, p. 133–136, févr. 2009.
10. S. Kharraja, Outils d aide à la planification et l ordonnancement des plateaux médico-techniques , Saint-Etienne, Université Jean-Monne, 2003.
11. M. Persson et J. A. Persson, "Health economic modeling to support surgery management at a Swedish hospital", Omega, vol. 37, no 4, p. 853–863, août 2009.
12. E. Marcon, S. Kharraja, N. Smolski, B. Luquet, et J. P. Viale, "Determining the number of beds in the postanesthesia care unit, a computer simulation flow approach", Anesth. Analg., vol. 96, no 5, p. 1415–1423, table of contents, mai 2003.

Chapter 6
An Immune Memory and Negative Selection to Visualize Clinical Pathways from Electronic Health Record Data

Mouna Berquedich, Oulaid Kamach, Malek Masmoudi, and Laurent Deshayes

Abstract Clinical pathways indicate the applicable treatment order of interventions. In this paper we propose a data-driven methodology to extract common clinical pathways from patient-centric electronic health record (EHR) data. The analysis of patients records can lead to better understanding and condoling pathologies. The proposed algorithmic methodology consists of designing a system of control and analysis of patient records based on an analogy between the elements of the new EHRs and the biological immune systems. We use biological immunity to develop a set of models for structuring knowledge extracted from EHR and to make pathway analysis decisions. A specific analysis of the functional data led to the detection of several types of patients who share the same EHR information. This methodology demonstrates its ability to simultaneously process data and is able to provide information for understanding and identifying the path of patients as well as predicting the path of future patients.

M. Berquedich (✉)
Laboratory of Innovative Technologies (LTI), Abdelmalek Saâdi University, Tangier, Morocco

Innovation Lab for Operations (ILO), University Mohammed 6 Polytechnic, Ben Guerir, Morocco
e-mail: mouna.berquedich@um6p.ma

O. Kamach
Laboratory of Innovative Technologies (LTI), Abdelmalek Saâdi University, Tangier, Morocco

M. Masmoudi
University of Jean-Monnet, Faculty of Sciences and Technologies, Saint-Étienne, France
e-mail: malek.masmoudi@univ-st-etienne.fr

L. Deshayes
Innovation Lab for Operations (ILO), University Mohammed 6 Polytechnic, Ben Guerir, Morocco
e-mail: Laurent.Deshayes@um6p.ma

M. Masmoudi et al. (eds.), *Artificial Intelligence and Data Mining in Healthcare*,
https://doi.org/10.1007/978-3-030-45240-7_6

6.1 Introduction

The widespread adoption of electronic health records (EHR) offers an unprecedented opportunity to apply computer techniques to clinical and operational data [1, 2]. Electronic health records store a wealth of clinical data that can potentially improve the quality of care in clinical emergency departments [1]. This information is often lost in the multitude of data collected. In their brief meetings, emergency physicians are too often confronted with poorly organized information that is difficult to synthesize and use at the bedside [5–7]. As electronic health records evolve, the integration of computerized clinical decision support provides the opportunity for sorting, collecting, and presenting this information, intending to improve patient's care [8, 9]. Clinical decision-support technologies have demonstrated the ability to improve patient outcomes in a variety of health care settings [10, 11]. Among several organizations is the Center for Medicare and Medicaid Services, which will continue to develop in this area [12].

Efforts to improve emergency care and regulatory incentives for their adoption have contributed to the rapid development and implementation of these technologies. Given the complexity of socio-technical systems (such as emergencies) in which they are implemented [13, 14] clinical decision-support technologies may have negative consequences [15].

Within the current revolution in the development and use of clinical decision support in emergency services, it is essential that the implementation of these technologies is based upon the best clinical evidence available to help improve patient care [16]. Indeed, the development of these technologies could benefit from the established design principles in the areas of health and human factors engineering. These technologies span over simple heuristics as "5 rights" of decision support—the right information, the right person, the right format, via the right channel, at the right moment in the workflow [17]—to more comprehensive theories of human factors engineering as Parasuraman's model of human interactions with automation [18].

We conducted a systematic review of the scope and influence of clinical decision-support technologies integrated with electronic health records and implemented in emergency departments. After examining these results of the systematic review, we further discuss the gaps of current researches while proposing an immune memory and negative selection to visualize clinical pathways from electronic health-recorded data. Clinical pathways reflect the best evidence available in practice. So, they indicate the order of intervention of an extensively applicable means of treatment and testing purposes. We propose a data-driven clinical practice development methodology to extract common clinical pathways from electronic health records.

A better analysis of the patient's path, using data centered on him, can lead to better treatments for patients with different pathologies. An algorithmic methodology is proposed to manage this type of data on pathways, focusing mainly on hospital data.

In this chapter, we design a system for monitoring and analyzing patient records based on an analogy between the elements of the new EHRs and the biological immune systems. We rely on biological immunity to develop a set of models to structure knowledge about EHR decisions and pathway analysis. A specific analysis of the functional data made it possible to detect several types of patients sharing the same information on their EHRs. This methodology demonstrates its ability to simultaneously process data and provide information, which will help to understand and identify the path of patients, as well as predict the path of future patients.

6.2 What is an EHR?

Electronic health records (EHRs) are a digital collection of patient health information in an electronic format and increasingly used by many developing countries, indeed. These records can be shared via a well-connected network of different health care settings [5]. The EHR includes demographic and personal statistics as age, weight, billing, vital signs, current and past medical history, family history, medical and allergic history, immunization status, laboratory test results, and radiological images [6]. The EHR systems are designed to reduce the number of paper-based medical records [1]. Their goal is to store data accurately and make them much more readable in a digital format [8]. Indeed, digital health records reduce the risk of data replication because the files can be shared between different health care systems [5]. Hence, they can be modified and updated by hospital data entry-operators [11]. These features reduce the risk of losing paper records [2]. That is EHR programs directly benefit physicians, patients and, of course, the hospital's managing authority [1]. The EHR system can also be beneficial in population studies and effective in extracting medical data for examination of possible trends and long-term changes in a patient [4].

6.2.1 Benefits with the EHR

The EHR is an excellent tool for managing longer, tedious documents more efficiently. It dramatically reduces transcription, refilling, and storage costs [5]. The EHR enables patient management with improved and accurate reimbursement coding [4]. As the software contains all the information related to patients, it significantly reduces the occurrence of medical errors and also contributes to improving the health of the patient through better treatment of diseases [26]. Here we discuss the five major benefits of EHRs over paper-based medical records:

1. **Costs:** Initiating the EHR, the upfront costs are much higher due to the large and digital configuration of the computer network, but the costs will dramatically decrease over time [46]. While the manual storage of paper records requires more staff to manage and maintain paper records, it allows access and organization of countless documents whose costs increase considerably over time [46]. That

EHR also saves labor, time, and physical storage space. In effect, that reduces costs in the long run.

2. **Storage:** EHR can be stored in a secure cloud, allowing easier access when needed. However, paper medical records required large warehouses [4]. Paper documents do not only occupy space but are not environmentally friendly and tend to deteriorate with handling and time. Hence, the increase in storage costs.
3. **Security:** Security is the major concern of paper and electronic storage systems, and both of them are also vulnerable to security threats. If an installation electronically stores the records without any appropriate and effective security systems, they are thus vulnerable to be on hands of unauthorized persons [5]. In effect, the result can lead to the misuse of information. If the records are stored on paper, they may be lost, damaged, or stolen as a result of human error. Natural disasters such as fires or floods also play an important role in the security of health records.
4. **Access:** The accessibility of electronic health records has an advantage over paper records. Digital health records allow health professionals to access information instantly, anytime, and anywhere to make them more effective [8]. However, paper medical records to be shared with health professionals must be physically provided, scanned, or emailed. The consequence is a waste of time that increases costs.
5. **Readability and Accuracy:** EHRs are often written with standard abbreviations, making them more accurate and legible throughout the world, reducing the risk of confusion [11]. However, handwritten medical records can be difficult to read, which can contribute to medical errors. Medical records on paper do not allow health professionals to write all the necessary information.

6.2.2 Better Practice Management with the EHR

EHR improves the management of medical practice through integrated planning systems that directly link appointments with progress notes and automate coding as well as managing the claims. This platform smoothly handles requests about the patient's condition and manages patient-specific charts. In a report published by Jamoom et al. [12], in the "National Conference on Health Statistics," they demonstrated that over 79% of EHR systems' users operate much more efficiently, 82% of EHR's users save time while electronically sending prescriptions (electronic prescriptions), and 75% of EHR users receive lab results faster than usual. That is to say that EHRs have improved communication with other multidisciplinary physicians, labs, and between different hospitals, allowing for care service faster to patients. As it is a digital platform that offers online connectivity, doctors granted access to patient information at any time while assigning a task to service providers, including laboratories and other physicians [13]. The follow-up examinations are often parts of the best patient care.

Besides, the EHR enables coders to program a tool that contributes to rapid and reliable data-based examinations resulted from the images and test reports embedded within that tool as well as its easier accessibility. Further, such a test can be controlled by the versatile EHR system which prevents unnecessary duplication of medical tests. The best part of these electronic records is that they are integrated with national and international databases as well as disease registries. That is, it helps physicians to track the epidemiological status of the disease being treated and prepare for emergencies to efficiently store, process, query, and analyze medical data [5]. For example, the Ayasdi organization provides information to Mount Sinai Medical Center in the USA about the genetic sequencing of Escherichia coli (E. coli) [6]. These data are used to study bacterial resistance to antibiotics. It uses topological data analysis as a contemporary research methodology capable of understanding the characteristics of data [6]. In addition, there are different areas in the health care industry, including medical imaging [7, 9], patient genomics [10, 11], electronic health records [12, 13], unstructured textual data [14, 15], devices as well as log and sensor data [16, 17]. These areas can be managed by big data techniques and infrastructures [18], particularly in the area of healthcare management.

The health sector has invested heavily in advanced medical technologies to improve medical decision-making. These technologies have been the source of much research [1–3]. They aim to improve the quality and effectiveness of health services through health information technology (HIT). However, most studies have focused on the commercial aspects of the system [4]. The assumption is that HIT will improve medical processes and reduce costs by integrating patient data or making them readily available to physicians and medical staff. In this vein, several studies have shown that electronic health records (EHRs), including the widely used version of HIT, can improve physicians' performance and quality of care [5–8]. Physicians can draw valid and reasonable conclusions about medical treatment, despite the imperfect information they receive [9] and the lack of easy access to complete medical history. They can, however, make bad decisions. The medical history retrieved by the EHR allows physicians to have a much more complete view of the patient. Providing detailed information and care options can simplify decision-making in many medical situations [10].

It is believed that the significant use of HIT has plenty of benefits such as (1) improved quality, safety, and efficiency of care; (2) better engagement with patients and families; (3) improved coordination of care; (4) improved public health; and (5) greater privacy and security [3]. These benefits should help clarify the criticisms of HIT decisions. However, the impact of EHRs on very stressful environments such as emergency services often faces overcrowding and time constraints that can reduce these benefits. Particularly, overcrowding in emergency departments often leads to unsatisfactory clinical outcomes [11–15], misdiagnosis, imperfect documentation, and incorrect pharmacotherapy [16, 17], and mediocre EHR [18]. Although the availability of medical information is essential to the success of medical care [19–21], they have been explored to determine strategies for diagnostics, therapeutics, and planning resources [28–30] and also to measure the effectiveness of triage of patients to emergencies using medical information systems [31]. Many factors influence decision-making, including uncertainty and risk conditions [32], heuristic methods [33], and experience [34].

Studies on the influence of medical systems on decision-making, in real-time, have shown that clinicians want access to aggregated data in terms of demographic data, tests, procedures, and treatments, and in particular results of the diagnosis [35]. Physicians are required to make consistent decisions during their daily work: diagnosis, therapeutic interventions, and the participation of other physicians are some of their common actions. To manage information and decision-making, clinicians can use HIT [36]. The EHR stores patient information in an organized and accessible manner (unlike paper charts that can be dispersed and easily misplaced), but can sometimes hinder regular flow [37].

Ben-Assuli et al. [11, 38] analyzed the effects of using the EHR on medical decision-making. They have shown that their use can improve decision-making and increase medical efficacy. The use of management orientation strategies (such as the "plan, study, act" model) can improve the performance of medical staff and help overcome the challenges of implementation [39]. The confidence of decision-makers and the time taken to make these decisions are also underlying elements of the decision-making process. It has been shown that, in general, a clinical assistance system can effectively screen patients using a limited amount of information [28, 40, 41] and can offer both clinical and economic benefits. Physician confidence also has a significant impact, especially in stressful environments, where doctors must make decisions to save lives. Some scholars found that interactions with fellow physicians increased the confidence of decision-makers [42], probably because they provided additional information to support the decision. Additionally, exposure to information related to a specific action plan influences decisions by increasing the level of trust [43]. When a medical history is available and accessed, clinicians can have more confidence, allowing them to make more effective decisions with less uncertainty bias [44]. Although physicians report having a high level of confidence after receiving information [24, 45], they are exposed to additional information. Therapeutic options have been shown to increase decision complexity in several medical situations [10].

Concerning the implementation of integrated medical information systems, Black et al. [46] pointed out that the inexorable increase in national health expenditure and the desire to improve the quality of healthcare have led to widespread the adoption of HIT, but their results should be examined and studied.

Some studies have been conducted to define much better a theoretical framework for assessing the potential value and cost-effectiveness of HTI [38, 47, 48]. Specifically, Claxton et al. [49], Walker et al. [19], and Kapoor and Kleinbart [50] found that electronic health information exchange (EID) and interoperability among healthcare providers can reduce costs for health maintenance organizations (HMOs). An HMO is an organization that provides health care services to insured persons and puts them in touch with health care providers (such as hospitals, doctors, and community care) on a prepaid basis. This type of organized care has been found to reduce hospitalization costs for care [51] and increase patient satisfaction [52]. The medical care provided in such an environment is complete because the HMO has many facilities and services in most branches of medicine and many doctors and medical staff.

Technology acceptance is another critical factor that has attracted attention in empirical studies in the health sector [53, 54]. The results suggest that the integration of HIT into health facilities is not without pitfalls. Many factors appear to affect rejection, including computer resistance [55], incorrect measurement methods [56], and misleading expectations on the part of IT staff [57].

Two most important design's principles that have emerged in terms of increasing the acceptance are the identification of the user's needs and the integration of workflow [58]. An additional element of HIT is the specific need of health professionals from different fields [59]. For instance, if psychiatrists need additional information such as compliance or abuse, and other functions within the system as a wireless patient's monitoring [60]. Pediatricians, on the other hand, check the growth parameters and may need the system to calculate a development curve [61–63].

6.3 Current Study

The era of mega-data has begun because of the large, complex, and growing number of datasets produced by various sources such as the Internet of Things (IoT), government archives, medical records, multimedia, telephone logs, social networks, media, and other digital traces [1–3]. Mega-data is used to transform medical practice, inform decision-making, and modernize public policies [4]. The number of complex data from medicine and health care is growing rapidly with many and essential information. As a result, mega-data have infinite potential for efficiently storing, processing, querying, and analyzing medical data [5]. However, issues as security, privacy, efficiency analysis, and data quality are very important in medical applications and data collection. As a result, medical information reveals valuable, intellectual property and its uses are highly protected. That means that information management should not be limited to the established practices and laws, but also be it expanded to privacy expectations of subjects [19]. The review of the global disease network, using biological databases and electronic health record (EHR) data, has revealed that the two are powerful means for understanding the complexity of the relationships between diseases [21, 22]. In the study [23], the authors proposed to combine different powerful approaches to compare the structure and connectivity of disease networks between three different populations: Hispanic/Latin American, Caucasian, and African-American populations.

Further, the acquisition of data from complex sources and heterogeneous patients is another big data challenge in the health sector. These challenges include obtaining clinical notes, understanding them correctly, organizing medical imaging data, collecting biomarker data, and understanding genomic data in large quantities. These data may be useful in the clinic during patient evaluation.

Furthermore, various sensors provide access to data on the psychological, behavioral, and social characteristics of the patient [24]. A recent survey by Fang et al. [25] reviewed the current challenges, techniques, and future directions of computerized health informatics in big data. They summarized the challenges in four dimensions (V) of mega-data: volume, velocity, veracity, and variety. In another recent survey [5], the authors considered that the problems posed by the 6V—volume, velocity, veracity, variety, variability, and value—represented a difficulty, because of the complexity of big data.

However, recent technologies as artificial intelligence (AI) can help to solve complex problems. Artificial intelligence, in its broadest sense, would demonstrate the ability of a machine to perform tasks similar to human thinking [3]. Thus, AI has been used for computer systems with a more complex task execution capability than just programming [26]. Artificial intelligence also increases in the three dimensions "V" (volumes, velocities, and variety of data), likewise big data. It allows the learning process, the delegation of difficult motive recognition, and additional responsibilities at the respect of computer methods in case of large volumes of data. Indeed, it contributes to the speed of data by allowing them to associate quick decisions to make different choices. However, the problem of the variety is not only solved by parallelizing and spreading the problems, whereas it is mitigated by capturing, structuring, and understanding unstructured data using AI and different analytics [27]. In effect, AI and large data analytics could (re)define the health system with better performance, provide healthcare information, and improve global processes as two essential elements to improve productivity, efficiency, and effectiveness and quality of care as well [28]. All told, in order to enhance the performance of artificial intelligence, it is demonstrated by professionals that computational intelligence (CI) techniques adapt to medical data, while CI generally refers to the ability of a computer to learn a particular task, from experimental observations or data, which facilitates intelligent behavior in complex problems and changing environments [29]. Moreover, the CI techniques are classified using simple and hybrid methods. Simple methods refer to the studies that use a single machine learning technique (e.g., genetic algorithm, particle swarm optimization—PSO, artificial immune system—AIS, artificial system and neural network—ANN) as the main methods. Hybrid methods involve studies that use hybridization of two or more methods (i.e., neuro-fuzzy—NF) and the fuzzy support vector machine (FSVM). The example of Latifoglu et al. [30], who used only the artificial immune recognition system (AIRS), as the main technique for the diagnosis of atherosclerosis from Doppler signals of the carotid artery, is considered a unique classification method. Gu et al. [31], in their study of the hybrid classification, also used the FSVM technique in the classification of medical datasets.

Recently, the concept "patient journey" (or the trajectory of the patient) is the central part of each discussion on the evolution of the health system in France [20]. The patient's journey is defined by the French Ministry of Health as "a global structured network of continuous care for patients, as much closer as it is possible to their home" [20].

Previously, the path of the patient refers to the succession of the treatment steps of a patient in a given hospital [2]. It can be a sequence of procedures (clinical examination, biological assay, biopsy, and surgery), a sequence of clinical stages (inflammation, proliferation, and maturation phases), or a sequence of medical units (surgical unit, intensive care unit, conventional hospitalization, rehabilitation, and housing).

Thus, the concept of the patient's journey has several different scales: within a hospital, the succession of points of contact between a patient and the health system, various geographical divisions (for example, regions and departments), and the territorial scale introduced by the reform of the Hospital Group Territory (GHT) a regulation that requires public hospitals to group on a geographical basis [21].

Through several case studies, we have focused our attention on several of these scales: at the hospital, the scale was a succession of medical units forming the patient's stay, a patient journey between different hospitals of the same region, and finally all the stays of a patient in a hospital. The latter is seen through the prism of the information available in the databases of hospitals.

6.4 Overview of the System

The artificial immune system field, being inspired from the natural immune system, a number of antigens that trigger to protect the body against a pathogen, of several species, helps ambitious practitioners to develop systems that operate within environments facing similar constraints that natural immune system deals with biologically. In this context, De Castro and Timmis [65] define the AIS as "the adaptive systems, inspired by the theories of the immunology, as well as the functions, the principles, and the immune models, in order to be applied to the resolution of problem."

The main objective of our work is to develop an essential support tool for hospital decision-makers, to reinforce the quality of their decisions in the face of the massive flow of patients. The basic idea is to detect traces in the database and help the leaders, identifying the bad scenarios using AIS techniques, including negative and clonal selection. The analogy between the principle of the natural immune system and the problem proposed below (Table 6.1) has prompted us to develop our system.

6.4.1 Representation of the Self-Cell

The normal situations of EHRs are represented by autonomous cells. In this work, we propose a model capable of directing and organizing knowledge related to normal and stable EHR situations. This model has four characteristics in its structure. It is presented in Table 6.2.

$$SC = \{Dt_j, D_j, M_j, P_j\} \tag{6.1}$$

Table 6.1 Analogy between the principle of the natural immune system and the developed system [65]

Natural immune system	Artificial immune system applied in patient's pathways
Body	EHR
Self	Normal pathway of patient
Infected cell	Disturbed pathway
Non-self (antigen)	False pathway
Antibody	Control decisions
Lymphocyte (B)	Combination of control decision for detected disturbance
Affinity	Adequacy between the correction actions & disturbance
Memory cells	Database
Response strategy	Immune Memory Based Algorithm

Table 6.2 Summary of EHRs

Patient	Date	Purpose	Medication	Procedure	Diagnosis
1	xx/yy/2012	P1	M1	N/A	D1
1	xx/yy/2013	P2	M2	P1	D2
1	xx/yy/2015	P3	M3	N/A	D1
1	xx/yy/2018	P4	M4	N/A	D3

- Dt_j: Entrance date of patient j
- D_j: Diagnosis type of patient j
- M_j: Medication prescribed for j
- P_j: Medical procedure followed by j

6.4.2 Antigen Representation

The disturbance situation is defined as any type of event that may affect the patient's i journey. That is, we characterize that disturbance as attributes' vector that shows the pathways of infection. We represent it as it follows:

$$V_i = [\text{Date, Diagnosis, Visit\_purpose, Medication, Procedure}]$$

The formula "$el_k, k = 1, \ldots., Kl, l = 1, \ldots, L$" denotes a set of situations Kl of specific types of L. This organizes a "scenario" which takes place during the phase of the patient's medical visits.

In Table 6.3, the notion of event "CKD Stage 4" refers to the type of diagnosis, the word "Diuretics" indicates the type of medication and the "renal" corresponds to the ultrasound procedural type.

Table 6.3 Detail of the EHRs

Patient	Visit date	Visit purpose	Procedure	Medication	Diagnosis
1	24/9/2012	P1	N/A	ACE inhibitors	CKD stage 4; hypertension
1	5/2/2013	P2	Renal ultrasound	ACE inhibitors, diuretic	AKI, CKD stage 4; hypertension
1	3/1/2015	P3	N/A	ACE inhibitors, diuretic statin	CKD stage 4; hypertension
1	3/6/2018	P4	N/A	ACE inhibitors, diuretic statin	AKI, CKD stage 5; hypertension

For example, the details of patient record 1 in Table 6.2 are presented in Table 6.3. The medication, diagnosis, and procedure scenarios are coded as M_1, D_1, and P_1, respectively.

Our main goal is to define data that is consistent with the common patterns of clinical outcomes. To do this effectively, we integrate negative selection and immune memory to project unique visitor content. Here, each patient will be identified by a unique combination, which highlights the purpose of their visit and specifies the procedure, diagnosis, and medications needed to meet their needs. We use negative selection here for the reduction of multidimensional records of visits. In other words, the recordings of the visits can be represented by a sequence of visits sorted by dates. Each patient has only a single sequence, starting with his first medical visit and ending with the last, recorded during the establishment of the (EHR).

6.4.3 Representation of B-Cells

Control strategies are represented by "B-cells," which allow the recognition process and neutralize the antigen detected in the concerned sphere. B-cells aggregate with one or more decision controls (antibodies), which react whenever the disturbance occurs. Therefore, the system must create a B-cell identical to each detected antigen. Equation 6.2 illustrates the definition of independent B-cells. Receptors are a prerequisite, which has a structure similar to that of the antigen described below (see equation). The set of epitope receptors and antibodies is equal to the B-cells. According to the suggested model, the control decision can be either 0 or 1. When the antibody is activated, the similar epitope indicates the value 1, and when it is disabled, the value 0 is assigned. Epitopes are the activators of antibodies in these cases.

$$\text{B-cell} = \{\text{Receptors},\ \text{Antibodies},\ \text{Epitopes}\} \tag{6.2}$$

6.5 Negative Selection Algorithm for System Monitoring

Disturbance detections are considered the first steps in the disturbance management process. Through this work, we focus on the adaptation of the negative selection algorithm to detect and monitor disturbances occurring in health facilities. The initial concepts and theory of negative selection were presented by Forrest et al. in 1994 [66]. We build our algorithm to differentiate between self and non-self-cells. At the same time, auto-cells are those that represent normal situations within the database. The set S involves auto-cells that represent an acceptable level of performance. That set must be well specified by the expertise of the professionals. Besides, the set of traces R defines the non-self-representative cells, which represent the erroneous pathways of the patients. The set R is initially empty. It must be filled by cells applying basic learning simulation mechanisms. These mechanisms will be detailed in the following.

6.5.1 *Step 1: Learning*

This step aims to produce sets of non-self cells, using periodic comparisons between normal situations (self-cells) and the state of the new EHR. We highlight matching ratios, to quantify existing distances in given situations, using data collected from history. These measures designated by SC and the elements of set S will be subjected to a similarity test. The matching rate is calculated mathematically by adapting Eq. (6.3).

$$Mat(V_j, SE) = \frac{100}{5} \sum_{i=1}^{5} \alpha_i \tag{6.3}$$

$Mat(V_j, SE)$ represents the corresponding percentage (%). The V_j indicates a normal situation among the set S. SE determines a new situation. Attribute values are extracted from the database α_{ij} and are calculated as follows:

$$\alpha_{ij} = \begin{cases} 1 & if \quad SE_i = V_{ij} \\ 0 & else \end{cases} \tag{6.4}$$

The SE_i is the ith attribute of an SE situation captured from the database (Fig. 6.1). The V_{ij} is the ith attribute of the jth situation (V_j) extracted from the set S. When the adequacy rate is lower than the coverage already fixed 1, the solution is, therefore, classified as abnormal and will then be added to the set R of the reasons. To close this process, the set R, which includes various types of erroneous paths, displays the most significant abnormal events. It is used in the next steps to identify the disturbances that occurred during the course of a patient.

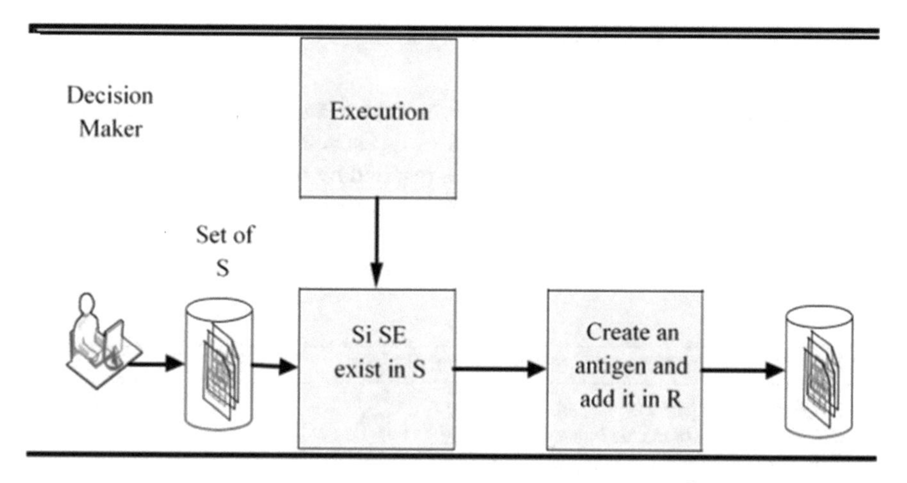

Fig. 6.1 Architecture based on the negative selection algorithm (NSA)

The algorithm is given as follows:

Algorithm 1: The negative selection algorithm

Data: the set of pathogens R
Result: Solutions
Initialization of a set S;
while *change* **do**
    change <– false;
    for *each Pathogen* Pg_j *of the set R* **do**
        Find similar pathogens to Pg_j in the set S using equation (3)
        if *no similar pathogen to* Pg_j *is detected* **then**
            Add the pathogen Pg_j to the set S;
            Change <– true;
            Delete the pathogen Pg_j from the set *R*;

6.5.2 Step 2: Monitoring

Given R is representing all the patterns built in the previous step. The T is designed to progressively read values from databases and measure similarities with the situation of R-set abnormalities (non-auto-cellular cells). Equation (6.5) can be used as a determinant of the matching rate, which groups all the abnormal situations of the set R with a current situation. Whenever the actual situation is the same as the already coded in the R-set, corresponding to a matching rate much higher than the fixed interval, it follows that a disturbance is identified and an antigen-specific is created. The reaction and response will then be activated by the T cells.

6.6 Control of EHRs by Memory Cells

The suggested methodology uses the NSA presented in the previous section to monitor pathways and detect disturbances. We suggest a second immune mechanism, called immune memory algorithm (IMA), inspired by biological immune memory to respond to disturbances during the patient's course. We present below a part of the pseudo algorithm of the immune memory.

CT0 and NP0 are initially in knowledge database.

Algorithm 2: Immune memory algorithm (IMA)

Data: P: a set of pathway, C: case of pathway
Result: CT: a set of cleaned traces of pathway
CT0: a set of cleaned traces of pathway at t0
NP0: set of P at t0
ND: Nearest distance
NP: set of nearest P
for *trace* $p \in P$ **do**
    Ct: case of trace in CT0 D: d(Ct, C)=|C-Ct| (D the Manhattan distance)
    ND = Min D
    We keep the smallest distance compared to a new pathway
    NP= NP0 U Np
    CT= CT0 U Ct

Where D presents the Manhattan distance, ND the nearest distance, and NP the set of nearest P.

6.6.1 The Algorithm Developed by Immune Memory (IMA)

Immune memory is a mechanism for selecting the best solutions. We have exploited the IMA in this study to detect and predict the good course for new patients. It works in four stages:

- **Step 1**—Data collection: the control algorithm acquires measurement data from the old patients database. In this case, the measurement data are linked to the number of patients in the database depending on the vector V.
- **Step 2**—Data analysis: The algorithm analyzes the data collected for possible disturbances, based on the negative selection algorithm presented in the previous section.
- **Step 3**—Selection of the pathway decision or appropriate trajectory: The algorithm selects from the database (immune memory) the correct path (cell B) capable of handling the perturbation (antigen Ag). For this, the algorithm determines the affinity between the detected disturbance (antigen) and all the

n paths (cells B) of the database. The algorithm selects the control decision with the highest affinity value. Affinity refers to the degree of similarity between the receptor element of a B-cell and the pathogen. The affinity measure considered in this paper is the distance from Manhattan [64], calculated using the following equation:

$$Affinity(Bcell, Ag) = \frac{\sum_{i=1}^{n} \sigma_i}{Length(Ag)} \tag{6.5}$$

where

$Length(Ag)$: length of antigen vector Ag

$\sigma_i = 1$ if $Bcell.receptors[i] = Ag.attribute[i]$, 0 else

- **Step 4**—The proposal of the decision of the best clinical course, to be followed by the patient concerned and validated by the hospital decision-maker.

6.7 Implementation and Results

This section presents the implementation of the EHR management system and discusses the results obtained. The proposed decision support system is implemented with the JAVA programming language. To evaluate the system, a hospital database, with a history of four years and given more than 100, 000 EHR patients, was deployed. We executed the AIS programs as it is presented in Algorithm 2. To construct the S set, we used a real patient records database (EHR). The above results show the phase of the regrouping of the relevant information in the form of a vector of five components. An analysis is then conducted according to two immune processes to communicate to the hospital decision-maker, the path closest to the path of the patient who entered within. The execution phase included a grouping and vector standardization compatible with the input patient information. Subsequently, two immune algorithms as previously explained: the negative selection and immune memory for the display of the optimal solution that responds to the patient input. The goal has been achieved by offering the hospital decision-maker a tool for detecting the patient's journey in broad and unstructured information. The most important is the fact of constantly feeding the database R, containing the erroneous solutions to avoid the medical errors of the old patients during their medical journey to the recent patients who share the same pathology and symptoms with them. Below are some results of the decision support tool produced:

In this example we show a sample of the results of the EHRs of the patients filtered by the negative selection algorithm and their execution time. Each EHR of a patient is presented as a Vi vector as already explained in the previous section; the following results show the phase of the regrouping of the relevant information in the form of a five-component vector; thereafter, an analysis is conducted according to two immune processes to communicate to the hospital decision-maker the path closest to the path of the patient entered. The execution phase included a grouping and vector standardization compatible with the input patient information. Consequently, the two algorithms already explained make it possible to give the optimal solution, in relation to the new case (i.e., new patient). The goal has been achieved, by offering the hospital decision-maker a tool for detecting the patient's journey, in broad and unstructured information. The most important is the fact of constantly feeding the database R, containing the erroneous solutions in order to avoid medical errors to recent patients who share the same pathology and symptoms with them.

As it is illustrated in Fig. 6.2, we realized an effective and efficient decision support system, which allows the filtering and the analysis of the data of the EHR of the different patients, by adopting the immune principles of the memory cells and the selection negative.

```
STARTING PHASE >
CLEANING PHASE >>
EHR ENTRY VECTORS: [ EHR [id=null, patient Id=-1, date=Tue Sep 25 21:59:42 WEST 2018, purpose=P10, procedure=P1,
medication=M1, diagnosis=D5],
EHR [id=null, patient Id=-1, date=Tue June 25 21:59:42 WEST 2018, purpose=P3, procedure=P2, medication=M8,
diagnosis=D5],
OPTIMAL SOLUTIONS (DISTANCE ORDER): {
80.0%=
Patient [id=1, nom=N_ABCDEFGHIJ, prenom=P_R, adresse=ADRESSE, dateNaissance=2018-09-24 22:21:33.0,
ehrs=[
EHR [id=1, patient Id=1, date=2018-09-24 22:21:33.0, purpose=P10, procedure=P6, medication=M1, diagnosis=D5],
EHR [id=2, patient Id=1, date=2018-09-24 22:21:33.0, purpose=P3, procedure=P2, medication=M8, diagnosis=D5], EHR [id=3,
patientId=1, date=2018-09-24 22:21:33.0, purpose=P2, procedure=P10, medication=M2, diagnosis=D7],
EHR [id=4, patient Id=1, date=2018-09-24 22:21:33.0, purpose=P2, procedure=P6, medication=M5, diagnosis=D2],
EHR [id=5, patient Id=1, date=2018-09-24 22:21:33.0, purpose=P10, procedure=P1, medication=M3, diagnosis=D10],
EHR [id=6, patient Id=1, date=2018-09-24 22:21:33.0, purpose=P1, procedure=P3, medication=M6, diagnosis=D5],
EHR [id=7, patient Id=1, date=2018-09-24 22:21:33.0, purpose=P5, procedure=P4, medication=M8, diagnosis=D9]]],
40.0%=
Patient [id=8397, nom=N_ABCDEFGHIJKLMNOP, prenom=P_L, adresse=ADRESSE, dateNaissance=2018-09-24 22:26:30.0,
ehrs=[
EHR [id=47771, patient Id=8397, date=2018-09-24 22:26:30.0, purpose=P10, procedure=P3, medication=M1, diagnosis=D5],
EHR [id=47772, patient Id=8397, date=2018-09-24 22:26:30.0, purpose=P8, procedure=P5, medication=M5, diagnosis=D8],
EHR [id=47773, patient Id=8397, date=2018-09-24 22:26:30.0, purpose=P4, procedure=P1, medication=M2, diagnosis=D4],
EHR [id=47774, patient Id=8397, date=2018-09-24 22:26:30.0, purpose=P1, procedure=P7, medication=M9, diagnosis=D8],
EHR [id=47775, patient Id=8397, date=2018-09-24 22:26:30.0, purpose=P10, procedure=P4, medication=M3,
diagnosis=D10]]],
20.0%=
Patient [id=9865, nom=N_ABCDEFGHIJKLMNOPQR, prenom=P_F, adresse=ADRESSE, dateNaissance=2018-09-24 22:28:18.0,
ehrs=[
EHR [id=55988, patient Id=9865, date=2018-09-24 22:28:18.0, purpose=P7, procedure=P4, medication=M6, diagnosis=D4],
EHR [id=55989, patientId=9865, date=2018-09-24 22:28:18.0, purpose=P3, procedure=P7, medication=M8, diagnosis=D10],
EHR [id=55990, patient Id=9865, date=2018-09-24 22:28:18.0, purpose=P9, procedure=P5, medication=M2, diagnosis=D9],
EHR [id=55991, patient Id=9865, date=2018-09-24 22:28:18.0, purpose=P2, procedure=P4, medication=M5, diagnosis=D2]]]
}>
END PHASE >
LOAD TIME: 3971 ms
```

Fig. 6.2 The results of EHR selection by affinity performance, data vectors of each EHR, and relevant results of the selection

6.8 Conclusion

Clinical pathways reflect the best evidence available in practice. Indeed, they indicate the order of intervention of the most widely applicable treatments for purposes of treatment and analysis. We propose a practical methodology for the development of clinical pathways. It is based on electronic health record data, centered on the patient. It aims to extract common clinical pathways, from data consistent with the clinical flow and facilitating patient care. A better analysis of the patient's path can lead to better treatments for patients with different pathologies. An algorithmic methodology is proposed that can handle this type of path data, focusing mainly on hospital data.

In this chapter, we design a patient record analysis system, based on an analogy between the elements of the new EHRs and the biological immune systems. Patient profiles are detected by the memory cells. We rely on biological immunity to develop a set of models for structuring knowledge about EHR and pathway analysis decisions. A specific analysis of functional data led to the detection of several types of patients who share the same information in their EHR. The developed system is based on a set of concepts and immune mechanisms. The negative selection algorithm is used to monitor the immune memory system and algorithm to select the appropriate pathway detection strategy to respond to the detected disturbances. The combination of negative selection mechanisms and immune memory gives the system the ability to recognize disturbances and select appropriate decisions.

This methodology demonstrated its ability to process data simultaneously. It can provide information for much better understanding and identification of patients' pathways as well as the prediction of the path of future patients.

References

1. C. Lovis, M. Ball, C. Boyer, P.L. Elkin, K. Ishikawa, C. Jaffe., Hospital and health information systems-current perspectives, Yearb. Med. Inform. (2011) 73–82.
2. P.G. Goldschmidt, HIT and MIS: implications of health information technology and medical information systems, Commun. ACM 48 (2005) 68–74.
3. E. Borycki, D. Newsham, D.W. Bates, Health in North America, Yearb. Med. Inform. 103 (2013) 6.
4. S.S. Jones, R.S. Rudin, T. Perry, P.G. Shekelle, Health information technology: an updated systematic review with a focus on meaningful use, Ann. Int. Med. 160 (2014) 48–54.
5. D.G. Goldberg, A.J. Kuzel, L.B. Feng, J.P. DeShazo, L.E. Love, EHRs in primary care practices: benefits, challenges and successful strategies, Am. J. Manage. Care. 18 (2012) e48–e54. Fig. 1. The demographics questionnaire screen in the simulation study (the first screen). 38 O. Ben-Assuli et al. / Journal of Biomedical Informatics 55 (2015) 31–40
6. A. Takian, A. Sheikh, N. Barber, We are bitter, but we are better off: case study of the implementation of an electronic health record system into a mental health hospital in England, BMC Health Serv. Res. 12 (2012) 484.

7. B. Jarvis, T. Johnson, P. Butler, K. O'Shaughnessy, F. Fullam, L. Tran, et al., Assessing the impact of electronic health records as an enabler of hospital quality and patient satisfaction, Acad. Med.: J. Assoc. Am. Med. Colleges 88 (2013) 1471–1477.
8. R.D. Cebul, T.E. Love, A.K. Jain, C.J. Hebert, Electronic health records and quality of diabetes care, N Engl. J. Med. 365 (2011) 825–833.
9. H.S. Sox, M.A. Blatt, M.C. Higgins, K.I. Marton, Medical decision making, American College of Physicians, Philadelphia, Pennsylvania, 2007.
10. D.A. Redelmeier, E. Shafir, Medical decision making in situations that offer multiple alternatives, J. Am. Med. Assoc. 273 (1995) 302–305.
11. O. Ben-Assuli, M. Leshno, I. Shabtai, Using electronic medical record systems for admission decisions in emergency departments: examining the crowdedness effect, J. Med. Syst. 36 (2012) 3795–3803.
12. D.W. Spaite, F. Bartholomeaux, J. Guisto, E. Lindberg, B. Hull, A. Eyherabide, et al., Rapid process redesign in a university-based emergency department: decreasing waiting time intervals and improving patient satisfaction, Ann. Emerg. Med. 39 (2002) 168–177.
13. S. Schneider, F. Zwemer, A. Doniger, R. Dick, T. Czapranski, E. Davis, Rochester, New York: a decade of emergency department overcrowding, Acad. Emerg. Med. 8 (2001) 1044–1050.
14. J.B. McCabe, Emergency department overcrowding: a national crisis, Acad. Med. 76 (2001) 672–674.
15. A. Bair, W. Song, Y.C. Chen, B. Morris, The impact of inpatient boarding on ED efficiency: a discrete-event simulation study, J. Med. Syst. 34 (2010) 919–929.
16. J. Fordyce, F.S.J. Blank, P. Pekow, H.A. Smithline, G. Ritter, S. Gehlbach, et al., Errors in a busy emergency department, Ann. Emerg. Med. 42 (2003) 324–333.
17. F. Lecky, J. Benger, S. Mason, P. Cameron, C. Walsh, G. Bodiwala, et al., The international federation for emergency medicine framework for quality and safety in the emergency department, Emerg. Med. J. (2013).
18. A.E. Lawson, E.S. Daniel, Inferences of clinical diagnostic reasoning and diagnostic error, J. Biomed. Inform. 44 (2011) 402–412.
19. J. Walker, E. Pan, D. Johnston, J. Adler-Milstein, D.W. Bates, B. Middleton, The value of health care information exchange and interoperability, Health Aff. 24 (2005) 5–10.
20. G. Hripcsak, S. Sengupta, A. Wilcox, R. Green, Emergency department access to a longitudinal medical record, JAMIA 14 (2007) 235–238.
21. R.D. Goldman, D. Crum, R. Bromberg, A. Rogovik, J.C. Langer, Analgesia administration for acute abdominal pain in the pediatric emergency department, Pediatr. Emerg. Care 22 (2006) 18–21.
22. W.R. Hersh, D.H. Hickam, How well do physicians use electronic information retrieval systems? A framework for investigation and systematic review, JAMA 280 (1998) 1347–1352.
23. J.I. Westbrook, E.W. Coiera, A.S. Gosling, Do online information retrieval systems help experienced clinicians answer clinical questions?, J Am. Med. Inform. Assoc. 12 (2005) 315–321.
24. J.I. Westbrook, A.S. Gosling, E.W. Coiera, The impact of an online evidence system on confidence in decision making in a controlled setting, Med. Decis. Making 25 (2005) 178–185.
25. W. Hersh, M. Helfand, J. Wallace, D. Kraemer, P. Patterson, S. Shapiro, et al., A systematic review of the efficacy of telemedicine for making diagnostic and management decisions, J. Telemed. Telecare 8 (2002) 197–209.
26. D. Blumenthal, M. Tavenner, The, "meaningful use" regulation for electronic health records, N Engl. J. Med. 363 (2010) 501–504.
27. H. Laerum, T.H. Karlsen, A. Faxvaag, Effects of scanning and eliminating paper-based medical records on hospital physicians' clinical work practice, JAMIA 10 (2003) 588–595.
28. M.E. Johnston, K.B. Langton, R.B. Haynes, A. Mathieu, Effects of computer-based clinical decision support systems on clinician performance and patient outcome: a critical appraisal of research, Ann. Int. Med. 120 (1994) 135–142.

29. V.L. Smith-Daniels, S.B. Schweikhart, D.E. Smith-Daniels, Capacity management in health care services: review and future research directions, Decision Sci. 19 (1988) 889–919.
30. K.K. Sinha, E.J. Kohnke, Health care supply chain design: toward linking the development and delivery of care globally, Decision Sci. 40 (2009) 197–212.
31. W. Michalowski, M. Kersten, S. Wilk, R. Słowinski, Designing man-machine interactions for mobile clinical systems: MET triage support using Palm handhelds, EJOR 177 (2007) 1409–1417.
32. G. Xue, Z. Lu, I.P. Levin, A. Bechara, The impact of prior risk experiences on subsequent risky decision-making: the role of the insula, NeuroImage 50 (2010) 709–716.
33. G. Gigerenzer, W. Gaissmaier, Heuristic decision making, Annu. Rev. Psychol. 62 (2011) 451–482.
34. S.F. Wainwright, K.F. Shepard, L.B. Harman, J. Stephens, Factors that influence the clinical decision making of novice and experienced physical therapists, Phys. Ther. 91 (2011) 87–101.
35. R.A. Greenes, E.H. Shortliffe, Commentary: Informatics in biomedicine and health care, Acad. Med. 84 (2009) 818–820.
36. M.A. Musen, B. Middleton, R.A. Greenes, Clinical decision – support systems, in: E.H. Shortliffe, J.J. Cimino (Eds.), Biomedical Informatics, Springer, London,(2014), pp. 643–674.
37. E. Dyer, V. Mohan, J.A. Gold, K.A. Artis, Data communication errors and the electronic health record in the intensive care unit: a pilot study, Am. J. Respir. Crit. Care Med. 189 (2014) 2803.
38. O. Ben-Assuli, M. Leshno, Using electronic medical records in admission decisions: a cost effectiveness analysis, Decision Sci. 44 (2013) 463–481.
39. A.S. McAlearney, J.L. Hefner, C. Sieck, M. Rizer, T.R. Huerta, Evidence-based management of ambulatory electronic health record system implementation: an assessment of conceptual support and qualitative evidence, Int. J. Med. Inform. 83 (2014) 484–494.
40. E.H. Shortliffe, Computer programs to support clinical decision making, JAMA 258 (1987) 61–66.
41. J. Wyatt, D. Spiegelhalter, Evaluating Medical Expert Systems: What to Test, and How? Knowledge Based Systems in Medicine: Methods, Applications and Evaluation, Springer, (1991) 274–90.
42. C. Heath, R. Gonzalez, Interaction with others increases decision confidence but not decision quality: evidence against information collection views of interactive decision making, Organ. Behav. Hum. Decis. Process. 61 (1995) 305–326.
43. A. Zylberberg, P. Barttfeld, M. Sigman, The construction of confidence in a perceptual decision, Front. Integr. Neurosci. 6 (2012) 1–10.
44. O. Ben-Assuli, I. Shabtai, M. Leshno, The impact of EHR and HIE on reducing avoidable admissions: controlling main differential diagnoses, BMC Med. Inform. Decis. Mak. 13 (2013) 49.
45. W.R. Hersh, M.K. Crabtree, D.H. Hickam, L. Sacherek, C.P. Friedman, P. Tidmarsh, et al., Factors associated with success in searching MEDLINE and applying evidence to answer clinical questions, JAMIA 9 (2002) 283–293.
46. A.D. Black, J. Car, C. Pagliari, C. Anandan, K. Cresswell, T. Bokun, et al., The impact of eHealth on the quality and safety of health care: a systematic overview, PLoS Med. 8 (2011) e1000387.
47. A. Basu, D. Meltzer, Value of information on preference heterogeneity and individualized care, Med. Decis. Making 27 (2007) 112–127.
48. H.E. Rippen, E.C. Pan, C. Russell, C.M. Byrne, E.K. Swift, Organizational framework for health information technology, Int. J. Med. Inform. 82 (2013) e1–e13.
49. K. Claxton, M. Sculpher, M. Drummond, A rational framework for decision making by the National Institute for Clinical Excellence (NICE), Lancet 360 (2002) 711–715.
50. B. Kapoor, M. Kleinbart, Building an integrated patient information system for a healthcare network, J. Cases Inform. Technol. 14 (2012) 27–41.
51. E.H. Wagner, T. Bledsoe, The rand health insurance experiment and HMOs, Med. Care 28 (1990) 191–200.

52. J. Schmittdiel, J.V. Selby, K. Grumbach, C.P. Quesenberry, Choice of a personal physician and patient satisfaction in a health maintenance organization, JAMA 278 (1997) 1596–1599.
53. L. Lapointe, S. Rivard, A multilevel model of resistance to information technology implementation, MIS Quart. 29 (2005) 461–491.
54. R.J. Holden, B.T. Karsh, The technology acceptance model: its past and its future in health care, J. Biomed. Inform. 43 (2010) 159–172.
55. C.E. Bartos, B.S. Butler, R.S. Crowley, Ranked levels of influence model: selecting influence techniques to minimize IT resistance, J. Biomed. Inform. 44 (2010) 497–504.
56. J.L.Y. Liu, J.C. Wyatt, The case for randomized controlled trials to assess the impact of clinical information systems, JAMA 18 (2011) 173–180.
57. M. Bloomrosen, J. Starren, N.M. Lorenzi, J.S. Ash, V.L. Patel, E.H. Shortliffe, Anticipating and addressing the unintended consequences of health IT and policy: a report from the AMIA 2009 health policy meeting, JAMIA 18 (2011) 82–90.
58. M. Peleg, J. Somekh, D. Dori, A methodology for eliciting and modeling exceptions, J. Boimed. Inform. 42 (2009) 736–747.
59. J.R. Vest, J. Jasperson, What should we measure? Conceptualizing usage in health information exchange, JAMIA 17 (2010) 302–307.
60. U. Varshney, Monitoring of Mental Health, Medication and Disability. Pervasive Healthcare Computing: EMR/EHR, Wireless and Health Monitoring. Georgia State University, Springer, Atlanta, GA, USA, 2009, pp. 259–280.
61. E. Borycki, R.S. Joe, B. Armstrong, P. Bellwood, R. Campbell, Educating health professionals about the electronic health record (EHR): removing the barriers to adoption, Knowledge Manage. E-Learn.: Int. J. 3 (2011) 51–62.
62. K.W. Hammond, E.N. Efthimiadis, C.R. Weir, P.J. Embi, S.M. Thielke, R.M. Laundry et al., Initial steps toward validating and measuring the quality of computerized provider documentation, in: AMIA Annual Symposium Proceedings: American Medical Informatics Association, 2010, p. 271.
63. X. Jing, S. Kay, T. Marley, N.R. Hardiker, J.J. Cimino, Incorporating personalized gene sequence variants, molecular genetics knowledge, and health knowledge into an EHR prototype based on the continuity of care record standard, J. Biomed. Inform. 45 (2012) 82–92.
64. L. N. De Castro, J. I. Timmis, Artificial immune systems: A new computational intelligence approach. Springer – Verlag, Berlin, 2002.
65. M. Berquedich, O. Kamach, M. Masmoudi, An Immune Memory and Negative Selection to Manage Tensions in Emergency Services. International Journal of Intelligent Engineering and Systems, 12 (2019) 214–228.
66. S. Forrest, A.S. Perelson, L. Allen, R. Cherukuri, Self-nonself discrimination in a computer, in: IEEE Symposium on Research in Security and Privacy, IEEE Computer Society Press, Los Alamos, CA, 1994.

Chapter 7
Optimized Medical Image Compression for Telemedicine Applications

Khalid M. Hosny, Asmaa M. Khalid, and Ehab R. Mohamed

Abstract Efficient compression of a huge number of medical images becomes necessary for storing and transmitting in telemedicine applications. In this paper, an algorithm is proposed for highly efficient compression of 2D medical images. The proposed algorithm used Legendre moments to extract the features from images and the whale optimization algorithm (WOA) to select which of these moments are the optimum to be used in the reconstruction process and in turn will produce the optimum reconstruction quality. The proposed algorithm aims to achieve higher compression ratios while maintaining the quality of the images. Medical images from different imaging modalities such as magnetic resonance imaging (MRI), computed tomography (CT), and X-ray images are used in testing the proposed algorithm. The mean square error (MSE), peak signal-to-noise ratio (PSNR), structural similarity index measure (SSIM), and normalized correlation coefficient (NCC) are quantitative measures used to evaluate the performance of the proposed algorithm and well-known existing medical image compression methods. The results showed that the quality of the reconstructed images using the proposed algorithm is much better than those of the conventional 2D compression algorithms in terms of MSE, PSNR, SSIM, and NCC.

7.1 Introduction

Medical images of different modalities such as magnetic resonance imaging (MRI), computed tomography (CT), and X-ray images are used to visually represent the internal structure of specific parts from the human body. These medical images are used by physicians for clinical purposes such as disease diagnoses and treatments.

Every day, hospitals and medical centers generate a very large number of medical images in order to review the physiological status of the patients [1]. Archiving these huge data requires very large storage capabilities. In addition, these medical

K. M. Hosny (✉) · A. M. Khalid · E. R. Mohamed
Zagazig University, Zagazig, Egypt

M. Masmoudi et al. (eds.), *Artificial Intelligence and Data Mining in Healthcare*,
https://doi.org/10.1007/978-3-030-45240-7_7

images need to be transmitted over networks for telemedicine applications [2]. In rural areas, where medical services may not be consistently available, the medical images of the patients are transmitted over different networks for consultation by medical experts. Sending/receiving a huge number of medical images via networks encounters transmission problems especially in the areas with limited bandwidth. To overcome these problems, efficient compression of medical images is essential in order to minimize the storage cost, retrieval time and limited bandwidth for image transmission [3]. Based on the nature of medical images where the fine details are important, compression of these images is constrained by preserving the fine details. Theoretically, a higher compression ratio means a reconstructed image with low quality and vice versa. Therefore, a successful compression method is the method which achieves a high compression ratio and produces a reconstructed image with high quality [4].

Generally, image compression can be classified into lossless and lossy compression. In lossless image compression, the original and reconstructed image are identical and the compression ratios are low. On the other side, high compression ratios are achieved by utilizing lossy compression methods with some information loss [5]. In medical images, the lossless compression is preferred in order to preserve the medical information of the image. However, many other applications, such as telemedicine and teleradiology, need high compression ratios in order to speed up the process of transmitting medical images across different networks, which can be achieved by lossy compression techniques.

7.2 Previous Approaches

In recent years, transform based medical images compression attracted researchers around the world. Discrete cosine transform (DCT) is one of the most popular transforms used in image compression due to its simplicity and efficiency. Wu et al. [6] proposed a DCT-based image compression algorithm which is based on adaptive sampling to decide the significant coefficients. A 2D medical image compression algorithm based on DCT and modified set partitioning in hierarchical trees (SPIHT) is proposed in [7].

Wavelet transform is another popular transform used in image compression because it captures both spatial and frequency information [5]. Haar wavelet transform as defined in [8] is utilized in medical image compression by Sahoo et al. [9] and Hasan and Harada [10]. Other wavelet-based 2D image compression algorithms are presented in [11, 12]. Wavelet transform has poor directionality and cannot represent edges and directions in 2D images [13]. To overcome this problem, many transforms have been proposed such as curvelet [14], contourlet [15], and ripplet transform [16].

The contourlet transform is based on a filter bank structure to deal with images with smooth contours. Contourlet-based image compression algorithms were proposed in [17, 18]. The curvelet transform can also represent 2D images with smooth curves [19]. Ripplet transform is a higher dimensional generalization of the curvelet transform. The ripplet transform was used in the compression of 2D medical images [20].

In the compression algorithms mentioned above, when trying to achieve high compression ratios, the image quality will be reduced and the fine details will be lost which is not preferred in medical images. However, the proposed compression algorithm can achieve high compression ratios (CR = 87.5%) with excellent reconstruction quality (MSE = 0.000488 and NCC = 0.9989). The reconstruction error is very close to zero and the reconstruction correlation is very close to 1 which is the best possible reconstruction quality.

Moment functions have been widely used in many image processing techniques such as image segmentation, image watermarking, pattern recognition, image reconstruction, and image compression. Block-based moment functions have been used successfully in image compression and reconstruction because they avoid numerical instabilities and result in good reconstruction quality. Block-based Tchebichef moments for bio-medical signal compression have been proposed as an alternative to DCT and other compression methods [21].

Pavlos et al. [22] proposed block-based Legendre moments image compression algorithm and compared the results with DCT compression, the results showed that the two transforms have similar compression properties, while DCT is slightly superior. Thung et al. [23] used Legendre and Tchebichef moments in medical image compression. They tested the robustness against different noises where their results showed that Legendre moments are more robust to noise than both DCT and Tchebichef moments.

In this paper, a new algorithm is proposed to compress the 2D medical images. In the proposed algorithm, the features of the input medical image are extracted using the block-based Legendre moments. The method of whale optimization algorithm (WOA) optimizes the performance of the compression algorithm by selecting the optimum features needed for image reconstruction which produce the optimum quality for the reconstructed image.

Numerical simulation is performed using a set of numerical experiments. These experiments were conducted with different medical imaging modalities such as MRI, CT, and X-ray. The obtained results show significant reconstructed images quality with high compression ratios.

The rest of this paper is organized as follows: Sect. 7.3 presents preliminaries about Legendre moments and the whale optimization algorithm. The proposed compression algorithm is described in Sect. 7.4. Performed experiments and their results are discussed in Sect. 7.5. The limitations of the proposed algorithm are discussed in Sect. 7.6. Finally, the conclusion is presented in Sect. 7.7.

7.3 Preliminaries

7.3.1 Legendre Moments

Legendre moments (LMs) are orthogonal moments defined in Cartesian coordinates and have an efficient representation of digital images [24]. The LMs are utilized in many applications such as template matching [25], content-based image retrieval [26], generating three-dimensional complex geological patterns [27], image segmentation [28, 29], and reconstruction of noisy medical image [30].

The 2D Legendre moments L of order $(p+q)$ for 2D image $f(x, y)$ are defined as follows [31]:

$$L_{pq} = \frac{(2p+1)(2q+1)}{4} \int_{-1}^{1} \int_{-1}^{1} P_p(x) P_q(y) f(x, y) d_x d_y, \tag{7.1}$$

where $p, q = 0, 1, 2, \ldots, \infty$ refer to the order of the Legendre polynomials P_p and P_q. The Legendre polynomials are complete set of orthogonal polynomials defined over the interval $[-1, 1]$. The pth order Legendre polynomial is defined as follows:

$$P_p(x) = \sum_{k=0}^{p} a_{k,p} x^k \tag{7.2}$$

where $a_{k,p}$ are the Legendre coefficients defined as follows:

$$a_{k,p} = (-1)^{(k-p)/2} \frac{1}{2^k} \frac{(k+p)!}{\left(\frac{(k-p)}{2}\right)!\left(\frac{(k+p)}{2}\right)!p!} \tag{7.3}$$

The Legendre polynomial obeys the following recurrence relation:

$$P_{p+1}(x) = \frac{(2p+1)}{(p+1)} x P_p(x) - \frac{p}{(p+1)} P_{p-1}(x), \tag{7.4}$$

with $P_0(x) = 1$, $P_1(x) = x$.

To calculate Legendre moments for an image, instead of using Eq. (7.1), matrix multiplication technique would be used for simplicity as follows:

1. According to Pavlos et al. [22], 8 × 8 Legendre kernel matrix can be calculated as follows:
   a. The roots of Legendre polynomial of order n (8 in our case because we divide the image into 8 × 8 blocks) are calculated, this will produce 8 roots.
   b. The value of Legendre polynomial P of order i (i= 1, 2, 3, ..., n) is calculated for each root value. This results in 8 × 8 matrix which represents the 8 × 8 Legendre kernel matrix.

The (8 × 8) Legendre kernel matrix is

$$K_{8\times 8} = \begin{bmatrix} 1.000 & -0.9600 & 0.8824 & -0.7718 & 0.6349 & -0.4796 & 0.3151 & -0.1506 \\ 1.000 & -0.7960 & 0.4504 & -0.0669 & -0.2446 & 0.4040 & -0.3857 & 0.2239 \\ 1.000 & -0.5250 & -0.0866 & 0.4257 & -0.3262 & -0.0323 & 0.3030 & -0.2677 \\ 1.000 & -0.1830 & -0.4498 & 0.2592 & 0.2543 & -0.2911 & -0.1143 & 0.2884 \\ 1.000 & 0.1830 & -0.4498 & -0.2592 & 0.2543 & 0.2911 & -0.1143 & -0.2884 \\ 1.000 & 0.5250 & -0.0866 & -0.4257 & -0.3262 & 0.0323 & 0.3030 & 0.2677 \\ 1.000 & 0.7960 & 0.4504 & 0.0669 & -0.2446 & -0.4040 & -0.3857 & -0.2239 \\ 1.000 & 0.9600 & 0.8824 & 0.7718 & 0.6349 & 0.4796 & 0.3151 & -0.1506 \end{bmatrix}.$$

2. The image is divided into blocks of size 8 pixels width and 8 pixels height (8 × 8 which is standard block size used in many compression algorithms such as JPEG and DCT [5–7]).
3. Legendre moments are calculated for each image block simply by matrix multiplication as follows:

$$M = K^{-1} B K^{-T}, \tag{7.5}$$

   where M refers to the calculated moments for image block B, K represents the Legendre kernel matrix, K^{-1} refers to the inverse of matrix K, and K^{-T} refers to the inverse of the transpose of matrix K.
4. The inverse Legendre moments can also be calculated by matrix multiplication as follows:

$$R = K M K^{T}, \tag{7.6}$$

   where R is the reconstructed image block.

7.3.2 *Whale Optimization Algorithm (WOA)*

WOA is a recent meta-heuristic optimization algorithm proposed by Mirjalili and Lewis [32] for continuous optimization problems. The WOA mimics the humpback whales hunting method that is called bubble-net hunting strategy. The humpback whales prefer to hunt school of krill or small fishes close to the surface, this foraging behavior is done by swimming around the prey within a shrinking circle and creating distinctive bubbles along a circle or "9"-shaped path. Before 2011, this behavior was only investigated based on the observation from the surface. However, in [33] it is found that this behavior utilizes tag sensors. For each bubble, two maneuvers are found and named "upward-spirals" and "double-loops." In the former maneuver, humpback whales dive around 12 m down and then start making bubble in a spiral shape around the prey while swimming up toward the surface. The latter maneuver includes three different stages: coral loop, lobtail, and capture loop. More detailed information can be found in [32–34].

7.3.2.1 Encircling Prey

Humpback whales can recognize the location of prey. In WOA, it is not possible to find the position of the optimum design in the search space a priori. So, WOA assumes that the target prey is the current optimum target solution or is close to the optimum. Once the best search agent is defined, the search agents (whales) update their positions according to the position of the best known solution (prey) as follows:

$$\overrightarrow{X}(t+1) = \overrightarrow{X^*}(t) - \overrightarrow{A}.D \tag{7.7}$$

$$D = \mid C.\overrightarrow{X^*}(t) - \overrightarrow{X}(t) \mid, \tag{7.8}$$

where t represents the current iteration, $\overrightarrow{X^*}(t)$ is the best position, $\overrightarrow{X}(t)$ is the current position, "." represents element-by-element multiplication, | | refers to absolute value, and A and C are coefficient vectors calculated by the following equations:

$$\overrightarrow{A} = 2\overrightarrow{a}.\overrightarrow{r} - \overrightarrow{a} \tag{7.9}$$

$$\overrightarrow{C} = 2.\overrightarrow{r}, \tag{7.10}$$

where $\overrightarrow{a}$ is linearly decreased from 2 to 0 over the course of iterations and r is a random vector in the interval [0, 1].

7.3.2.2 Bubble-Net Attacking Method (Exploitation Phase)

This phase can be modeled using two approaches as follows:

1. Shrinking encircling mechanism
   The shrinking encircling behavior is achieved by decreasing the value of $\overrightarrow{a}$ in Eq. (7.9). Note that the value of $\overrightarrow{A}$ is also decreased by $\overrightarrow{a}$. In other words, $\overrightarrow{A}$ refers to a random value in the interval [−a,a], where a is linearly decreased from 2 to 0 over the course of iterations. If $\overrightarrow{A}$ is a set of random values in [−1,1], the new position of a search agent can be defined somewhere in between the original position of the agent and the position of the current best agent.

2. Spiral updating position
   To mimic the helix-shaped movement of humpback whales, a spiral equation is used to update the position between the whale and the prey as follows:

$$\overrightarrow{X}(t+1) = D^{'}.e^{bl}.\cos(2\pi l) + \overrightarrow{X^{*}}(t), \quad (7.11)$$

   where $D^{'} = \mid \overrightarrow{X^{*}}(t) - \overrightarrow{X}(t+1) \mid$ indicates the distance of the ith whale and the prey (best solution), b is a constant for defining the shape of the logarithmic spiral, l is a random number in the interval [-1, 1], and the symbol e represents Euler's number.
   The two mechanisms, shrinking encircling and the spiral-shaped path, have equal probability of 50% during the optimization process as follows:

$$\overrightarrow{X}(t+1) = \begin{cases} \text{shrinking encircling, Eq. (7.7) if } p < 0.5, \\ \text{spiral shaped path, Eq. (7.11) if} P \geq 0.5, \end{cases} \quad (7.12)$$

   where p is a random number in the interval [0, 1].

7.3.2.3 Search for Prey (Exploration Phase)

In this phase, the position of the search agent is updated according to a randomly chosen search agent X_{rand}, this allows the WOA algorithm to perform a global search. This phase is implemented as follows:

$$D = \mid \overrightarrow{C}.\overrightarrow{X_{rand}} - \overrightarrow{X} \mid \quad (7.13)$$

$$\overrightarrow{X}(t+1) = \overrightarrow{X_{rand}} - \overrightarrow{A}.\overrightarrow{D}, \quad (7.14)$$

where X_{rand} is a random solution (random whale) from population of solutions. $\overrightarrow{A}$ is set a random value greater than 1 or less than -1 to force search agent to move far away from a reference whale.

The pseudocode of WOA is as follows:

Algorithm 1: Pseudocode of WOA

Data: the WOA control parameters as follows:
$MaxIter$: Maximum Number of Iterations.
Objective_function: MSE.
Result: (X^*) Optimum Solution according to the objective function.
Randomly initialize the whale population ;
Evaluate the objective function values and find the best search agent X^* ;
while $t < MaxIter$ **do**
    for *each search agent* **do**
        update a, A, C, l, and p ;
        if $p < 0.5$ **then**
            if $\mid A \mid < 1$ **then**
                update the position of the current search agent using Eq. (7.7) ;
            else
                select a random search agent X_{rand} update the position of the current search agent using Eq. (7.14) ;
            end
        else
            update the position of the current search agent using Eq. (7.11) ;
        end
    end
    calculate the objective function of each solution;
    update X^* if there is a better solution;
    $t = t + 1$;
end
return X^* ;

7.4 The Proposed Compression Method

In order to improve the quality of the reconstructed image, the optimum coefficients (Legendre moments) should be selected instead of the random selection of the coefficients. In the proposed algorithm, the selection of the coefficients is based on the optimization technique WOA where the WOA algorithm helps in selecting the optimum coefficients from the set of all possible coefficients. The mean square error (MSE) is the objective function of the WOA algorithm. The MSE measures the similarity between the original and reconstructed images where $MSE = 0$ is the ideal value. Figure 7.1 shows the framework of the proposed compression and decompression processes.

In the proposed algorithm, the LMs represent candidate solutions to the problem; the fitness of each solution is determined by minimizing the objective function (MSE). For predetermined number of iterations, the coefficients that achieve the

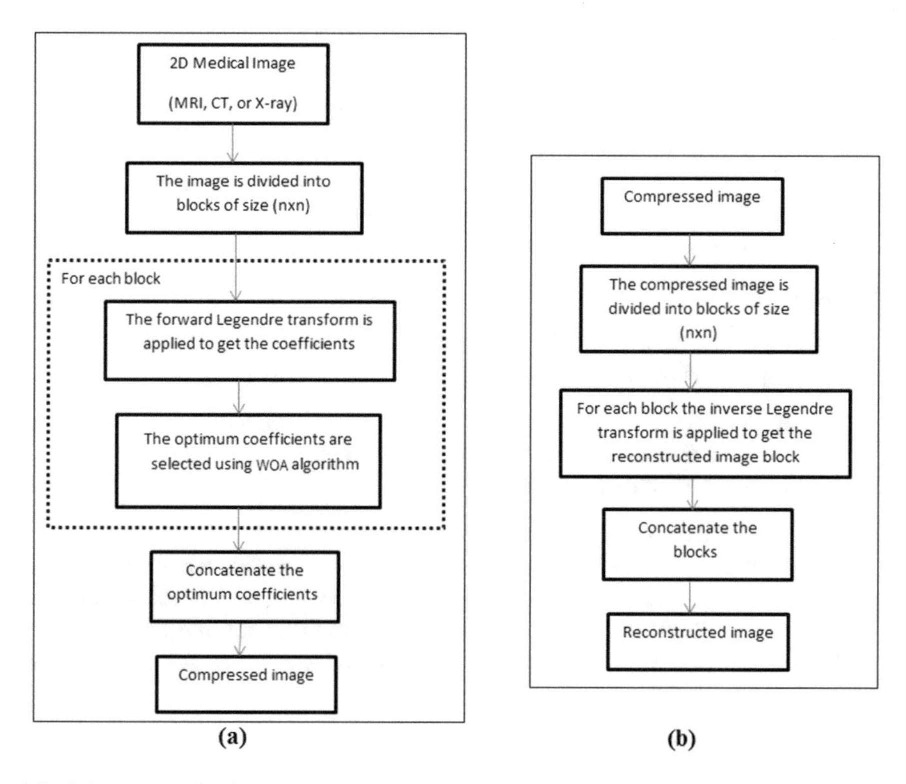

Fig. 7.1 Framework diagram of the proposed 2D medical image compression algorithm: (**a**) compression and (**b**) decompression

minimum MSE between the original and the reconstructed image represent the optimum coefficients. The proposed algorithm is summarized as follows: The input 2D medical image is divided into blocks of size n × n. According to the desired compression ratio (CR), the required number of coefficients (RNC) is calculated as follows:

$$\text{RNC} = \text{round}\left(\left(1 - \frac{\text{CR}}{100}\right) \times n^2\right) \tag{7.15}$$

Equation (7.5) is used for each block to compute LMs where MSE as optimization objective function of the WOA algorithm is applied to select the best moments where these selected moments are used in the decompression process. In the inverse process, the image is reconstructed using the inverse Legendre transform for each block using Eq. (7.6).

The proposed algorithm is summarized as:

1. Input the test medical image.
2. The input 2D medical image is divided into blocks of size $n \times n$.

3. Select the desired compression ratio (e.g., CR = 50%).
4. Based on the CR, the number (RNC) is calculated using Eq. (7.15).
5. The parameters of WOA algorithm are settled as follows:
   a. Dimension of the problem (D) = RNC;
   b. The number of solutions in population (NP) = 30 (as mentioned in [32]);
   c. Minimum value = 1; maximum value = $n \times n$;
   d. Maximum number of iterations (MaxIter) = 10,000.
6. For every block:
   a. Compute the LMs using the forward Legendre transform using Eq. (7.5).
   b. Apply the WOA algorithm to select the optimum LMs as follows:
      i. Initial population of solutions is randomly generated and the objective function (MSE) is calculated for each solution.
      ii. For each solution (x) in the population:
         - Update a, A, C, l, and p.
         - If $p < 0.5$, check the value of A, if $|A| < 1$, update the position of the current search agent using Eq. (7.7), if $|A| \geq 1$, select random solution and update the position of the current search agent using Eq. (7.14).
         - If $p \geq 0.5$, update the position of the current search agent using Eq. (7.11).
      iii. Calculate the objective function of each solution.
      iv. Update the best solution.
      v. Repeat steps (ii)–(iv) until the maximum number of iterations is reached.
   c. The coefficients computed using step (a) are classified into two classes. First, the optimum coefficients whose locations are the values in the best solution. Second, the other coefficients are set to zero.
   d. Concatenate the coefficients for each block to get the compressed image.
   e. The inverse Legendre transform is applied with all image blocks to reconstruct the image using Eq. (7.6).
7. The quantitative measures, mean square error (MSE), peak signal-to-noise ratio (PSNR), structural similarity index measure (SSIM), and normalized correlation coefficient (NCC) are used to evaluate the proposed compression algorithm (these quantitative measures are explained in detail in Sect. 7.5.2).

7.5 Numerical Experiments

7.5.1 *Test image*

The performance of the proposed compression algorithm is evaluated through a series of numerical experiments. A set of 2D medical images from different imaging modalities (MRI, CT, and X-ray) with different sizes (128×128 and 256×256) are selected from the open source library for medical images [35]. These images are displayed in Fig. 7.2. The selected images are defined using the digital imaging and communications in medicine (DICOM) file format. The DICOM is a standard file format developed by the National Electrical Manufacturers Association (NEMA) for storing, processing, and transmitting the medical images [36].

7.5.2 *Performance Measures*

The similarity between the reconstructed and the original medical images with different CRs is an explicit indicator for the success of the proposed compression algorithm. The evaluation measures are:

1. Compression ratio (CR %): is calculated as follows:

$$\mathrm{CR} = \left(1 - \frac{\text{size of compressed image}}{\text{size of original image}}\right) \times 100. \tag{7.16}$$

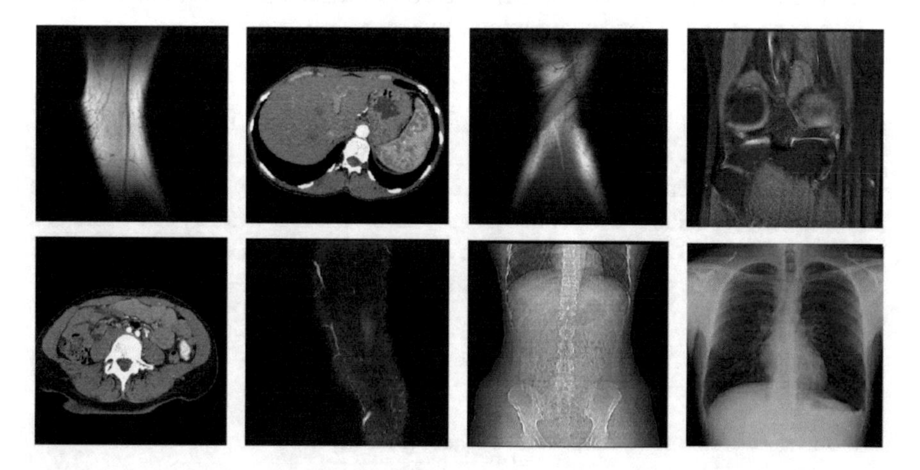

Fig. 7.2 Selected DICOM images for different medical imaging modalities

2. Mean square error (MSE): is used to measure the error between the original image, $f(x, y)$, and the reconstructed image, $F(x, y)$. It is calculated as follows:

$$\mathrm{MSE} = \frac{1}{M \times N} \times \sum_{x=0}^{M-1} \sum_{y=0}^{N-1} (f(x, y) - F(X, y))^2, \tag{7.17}$$

where $M \times N$ is the size of the 2D medical image.

3. Peak signal-to-noise ratio (PSNR): is used to measure the ratio between the maximum possible value (power) of the image pixel and the power of distorting noise as follows:

$$\mathrm{PSNR} = 20 \log_{10} \frac{255}{\mathrm{MSE}}. \tag{7.18}$$

4. Normalized correlation coefficient (NCC): is a measure used to determine the degree to which the two images are associated. The value of the correlation ranges from 1 to −1, where the number "1" indicates perfect positive correlation, and the number "−1" indicates perfect negative correlation. The NCC is calculated as follows:

$$\mathrm{NCC} = \frac{\sum_{x=0}^{M-1} \sum_{y=0}^{N-1} (f(x, y) \times F(X, y))}{\sqrt{\sum_{x=0}^{M-1} \sum_{y=0}^{N-1} (f(x, y) \times f(x, y)) \times \sum_{x=0}^{M-1} \sum_{y=0}^{N-1} (F(X, Y) \times F(X, Y))}}, \tag{7.19}$$

where f(x,y) and F(x,y) are the original and reconstructed images, respectively; $M \times N$ is the size of the input 2D medical image.

5. Structural similarity index (SSIM): is used to measure the structural similarity between the original and reconstructed images as follows:

$$\mathrm{SSIM}(x, y) = \frac{(2\mu_x \mu_y + C_1)(2\sigma_{xy} + C_2)}{(\mu_x^2 \mu_y^2 + C_1)(\sigma_x^2 \sigma_y^2 + C2)}, \tag{7.20}$$

where x and y refer to the original and reconstructed images, μ_x is the average value of x, μ_y is the average value of y, σ_x^2 is the variance of x, σ_y^2 is the variance of y, σ_{xy} is the covariance of x and y, and C_1 and C_2 are two variables to stabilize the division with weak denominator where

$$\sigma_x^2 = \frac{\sum_{i=1}^{N} (x(i) - \mu)^2}{N} \tag{7.21}$$

$$\sigma_y^2 = \frac{\sum_{i=1}^{N} (y(i) - \mu)^2}{N} \tag{7.22}$$

$$\sigma_{xy} = \frac{1}{N^2} \sum_{i-1}^{N} \sum_{j-1}^{N} \frac{1}{N} (x_i - x_j)(y_i - y_j) \tag{7.23}$$

$$C_1 = (k_1 L)^2 \tag{7.24}$$

$$C_2 = (k_2 L)^2, \tag{7.25}$$

where $k_1 = 0.01$, $k_2 = 0.03$ by default, and $L = 2^{\text{number of bits per pixel}} - 1$.

7.5.3 *Results and Discussion*

In this subsection, the proposed compression algorithm is implemented in MATLAB R2015a and tested using a large set of 2D medical images [35]. This paper shows the compression results of three sample images from the large dataset, Abdomen_CT.dcm of size 256 × 256, Leg_MRI.dcm of size 128 × 128, and Chest_Xray.dcm of size 128 × 128. The numerical simulation consists of two sets of numerical experiments. The first set of performed experiments aims to show the superiority of the WOA with Legendre moments over the conventional compression algorithm which uses Legendre moments only [23]. In the second set of the performed experiments, the performance of the proposed method is evaluated against the well-known existing methods.

The 2D medical images mentioned above are compressed by using both compression algorithms for different CRs (50%, 75%, and 87.5%). The evaluation measures, MSE, PSNR, SSIM, and NCC, are computed for each compressed medical image. The performance of both compression algorithms is evaluated in a quantitative form where the computed values of the evaluation measures are shown in Table 7.1. It is clear that the WOA optimization compression algorithm achieves an obvious improvement in compression with the conventional compression algorithm.

Since the visual perception plays an important role in image quality assessment, a comparison is made between the reconstructed images of the proposed algorithm and the reconstructed images of the conventional Legendre moments [23] with the CRs, 50%, 75%, and 87.5%. Figure 7.3 displayed the reconstructed Abdomen_CT.dcm image in two rows. The images in the upper row are reconstructed by using the conventional compression algorithm, while the images in the lower row are reconstructed by using the proposed compression algorithm. The medical images reconstructed by the proposed method are very close to the original ones where their visual quality is much better than the other images that are reconstructed by the conventional algorithm. On the other side, the quality difference between the original and conventional Legendre-based reconstructed medical images gets more noticeable at the higher CRs. Similar experiments are conducted with Leg_MRI.dcm and Chest_Xray.dcm medical images where consistent results are obtained and displayed in Figs. 7.4 and 7.5, respectively.

Table 7.1 The evaluation measures, MSE, PSNR, SSIM, and NCC, are computed using: (1) Legendre moments and (2) optimized Legendre moments using WOA

Algorithm		(1) Legendre moment [23]			(2) Optimized Legendre moment using WOA		
Image	CR	50%	75%	87.5%	50%	75%	87.5%
Abdomen_CT.dcm	MSE	0.0095	0.0680	0.2283	0.000734	0.0071	0.0288
(256 × 256)	PSNR	68.3375	59.8067	54.5454	78.1053	68.2443	61.6872
	SSIM	0.9425	0.8193	0.6838	0.9764	0.9230	0.8283
	NCC	0.9994	0.9958	0.9859	0.9999	0.9986	0.9972
Leg_MRI.dcm	MSE	0.0099	0.0975	0.2701	0.000874	0.0051	0.0255
(128 × 128)	PSNR	68.1677	58.2428	53.8150	77.2412	70.0442	64.2224
	SSIM	0.9219	0.7657	0.6154	0.9763	0.9404	0.8772
	NCC	0.9983	0.9836	0.9536	0.9989	0.9983	0.9965
Chest_Xray.dcm	MSE	0.000311	0.000682	0.0024	0.0000361	0.000225	0.000488
(128 × 128)	PSNR	83.1942	79.7924	74.3886	91.9589	85.1549	80.2336
	SSIM	0.9362	0.8649	0.7158	0.9833	0.9612	0.9159
	NCC	0.9993	0.9985	0.9947	0.9989	0.9987	0.9979

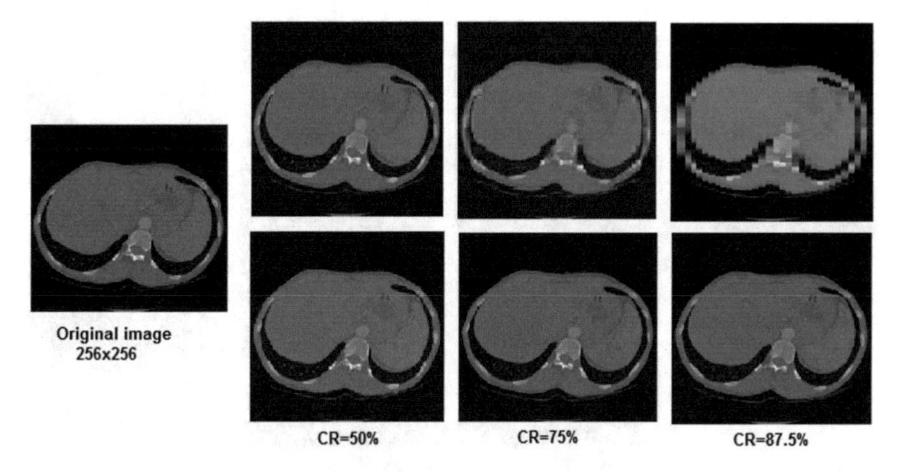

Fig. 7.3 The reconstructed Abdomen_CT.dcm medical images using the proposed and conventional Legendre compression algorithms for CR (50%, 75%, and 87.5%)

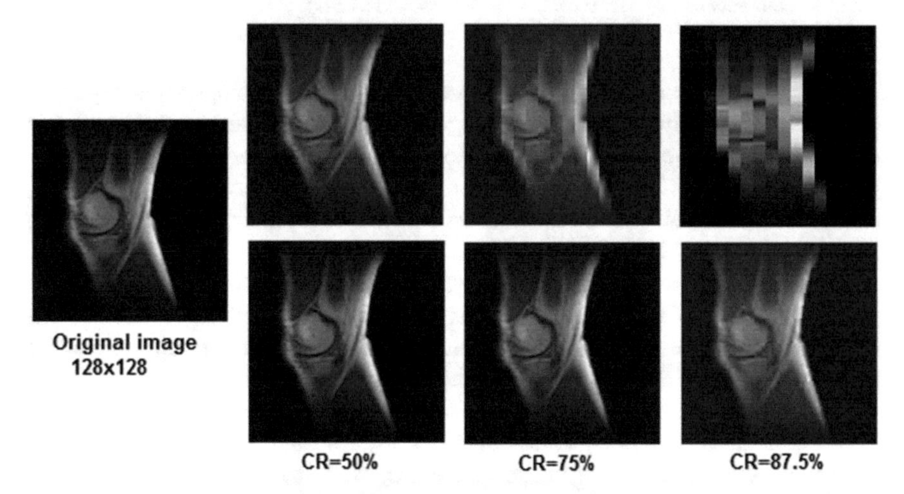

Fig. 7.4 The reconstructed Leg_MRI.dcm medical images using the proposed and conventional Legendre compression algorithms for CR (50%,75%, and 87.5%)

Also, a comparison is made between the proposed algorithm and the conventional Legendre moments in terms of the error (MSE) between the original and reconstructed images. The MSE is calculated for the three medical images mentioned above with the CRs, 50%, 75%, and 87.5%. The values of the MSEs are plotted and displayed in Fig. 7.6. By using the proposed algorithm, the values of MSE are very close to zero which means that the proposed algorithm minimized the reconstruction error that occurred in the compression–decompression processes.

A comparison with the well-known existing medical image compression algorithms is performed: DCT-based compression algorithm [7], Tchebichef moment-

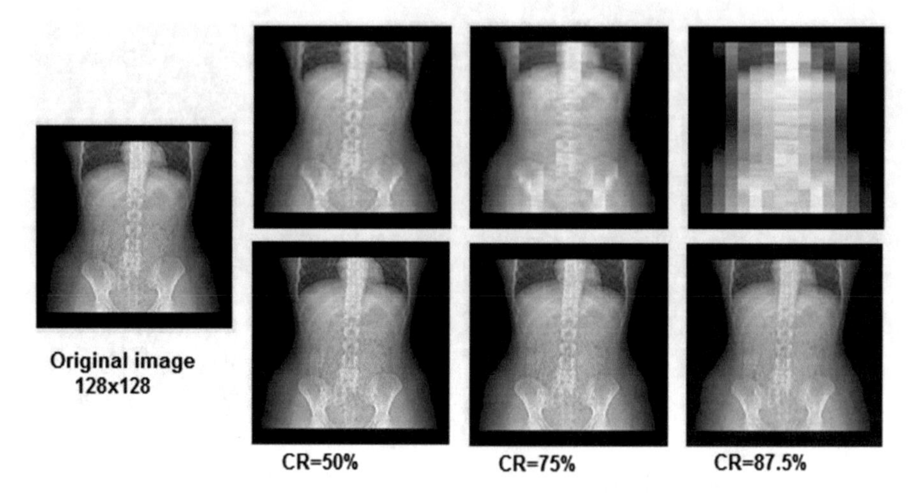

Fig. 7.5 The reconstructed Chest_Xray.dcm medical images using the proposed and conventional Legendre compression algorithms for CR (50%,75%, and 87.5%)

based compression algorithm [4], Legendre moment-based compression algorithm [23], Haar Wavelet-based compression algorithm [9], contourlet-based compression algorithm [18], curvelet-based compression algorithm [19], and ripplet-based compression algorithm [20]. The comparison is performed in terms of MSE, PSNR, SSIM, and NCC against CRs for three medical images of different medical imaging modalities Figs. 7.7, 7.8, 7.9, and 7.10.

In Fig. 7.7, the CR-MSE curve is shown and compared with existing compression algorithms. From the figure, we can see that the proposed algorithm has the minimum MSE over the other techniques. This means that the performance of the proposed algorithm exceeds the existing compression algorithms in terms of MSE.

In Fig. 7.8, the relationship between the CR and PSNR is plotted. The figure shows that the proposed algorithm achieves the maximum PSNR values which means that the reconstructed image of the proposed algorithm is more robust to noise in comparison with the other existing techniques.

In order to measure the similarity between the original and reconstructed images, the relationship between CR and SSIM is plotted in Fig. 7.9. From the figure, it is shown that the proposed algorithm has the highest SSIM values which means that the reconstructed images of the proposed algorithm are more similar to the original images than those of other existing techniques.

Finally, Fig. 7.10 shows a comparison between the proposed algorithm and other existing techniques in terms of the correlation (NCC) between the original and reconstructed images. It is shown from the figure that the proposed algorithm has the highest NCC values over the existing techniques. The overall results ensured the superiority of the proposed compression algorithm.

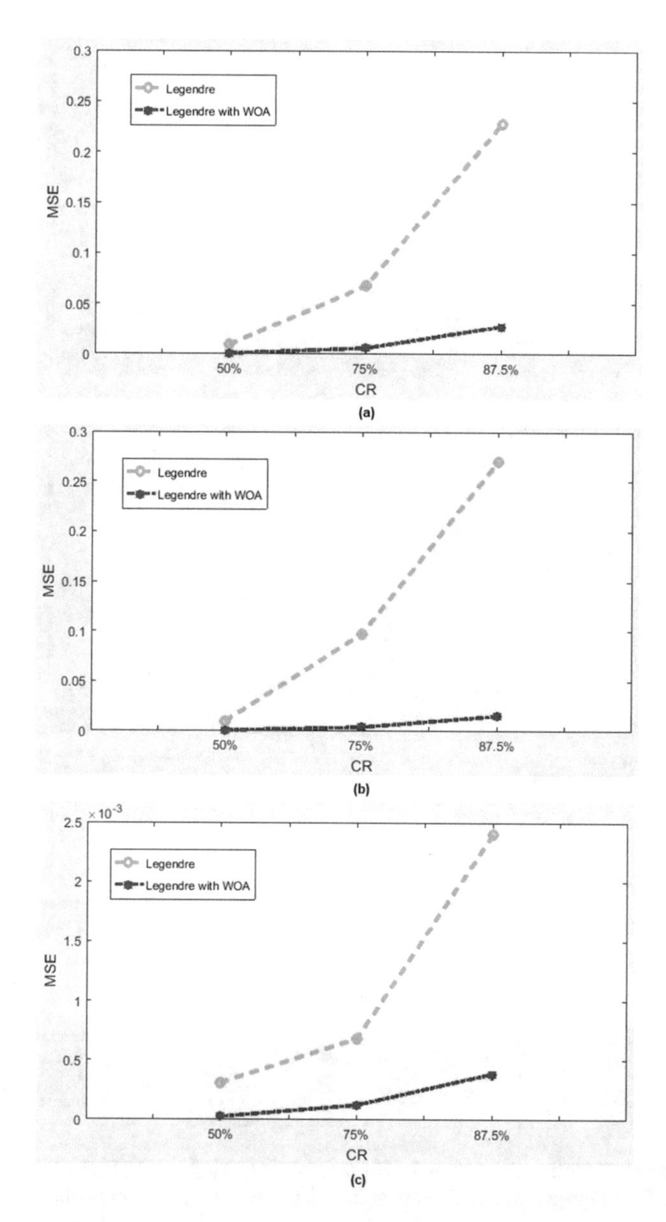

Fig. 7.6 The MSE for the proposed and the conventional compression algorithms: (**a**) Abdomen_CT.dcm image, (**b**) Leg_MRI.dcm image, and (**c**) Chest_Xray.dcm image

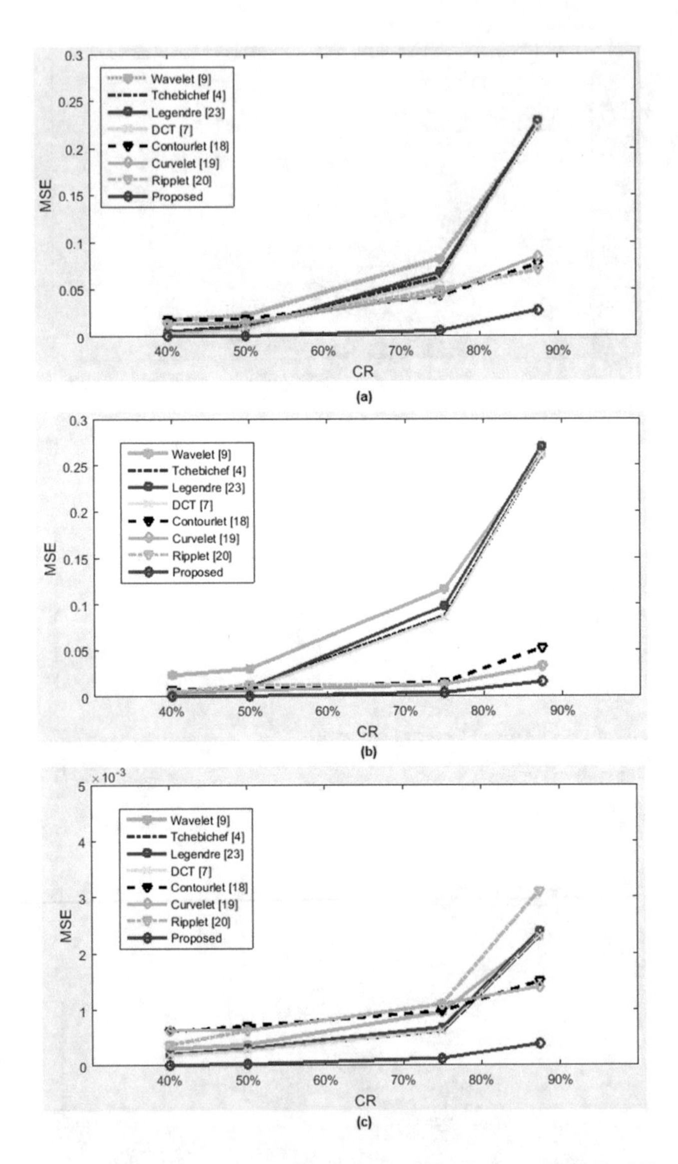

Fig. 7.7 The MSE for the proposed and other methods [4], [7], [9], [18], [19], [20], and [23]: (**a**) Abdomen_CT.dcm image, (**b**) Leg_MRI.dcm image, and (**c**) Chest_Xray.dcm image

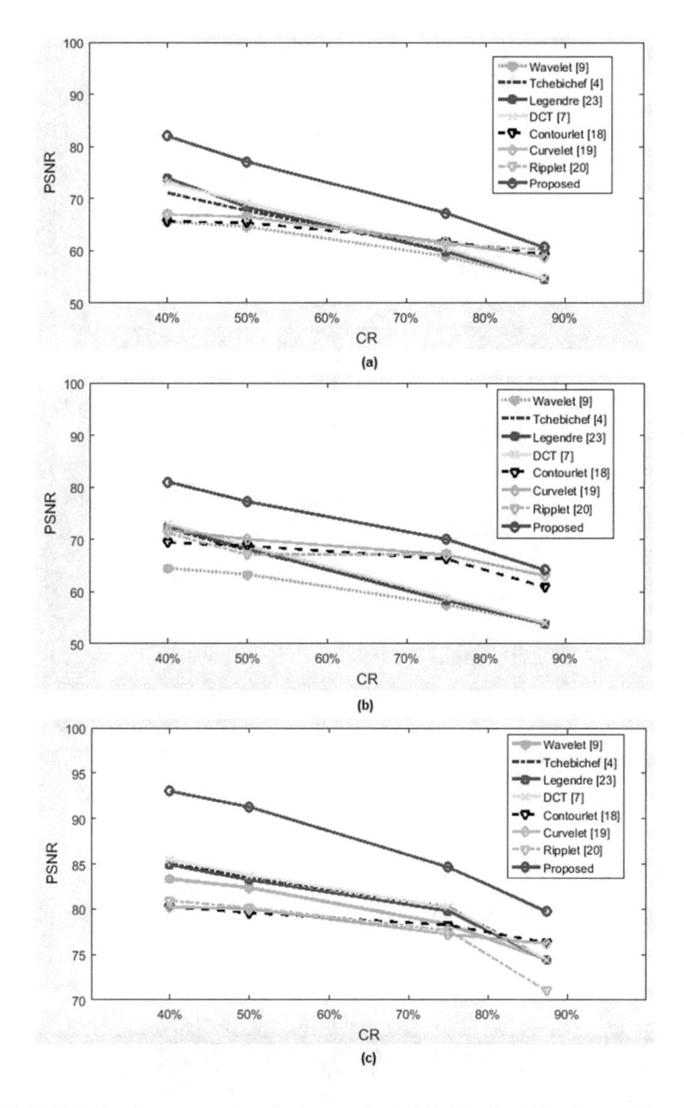

Fig. 7.8 The PSNR for the proposed and other methods [4], [7], [9], [18], [19], [20], and [23]: (**a**) Abdomen_CT.dcm image, (**b**) Leg_MRI.dcm image and (**c**) Chest_Xray.dcm image

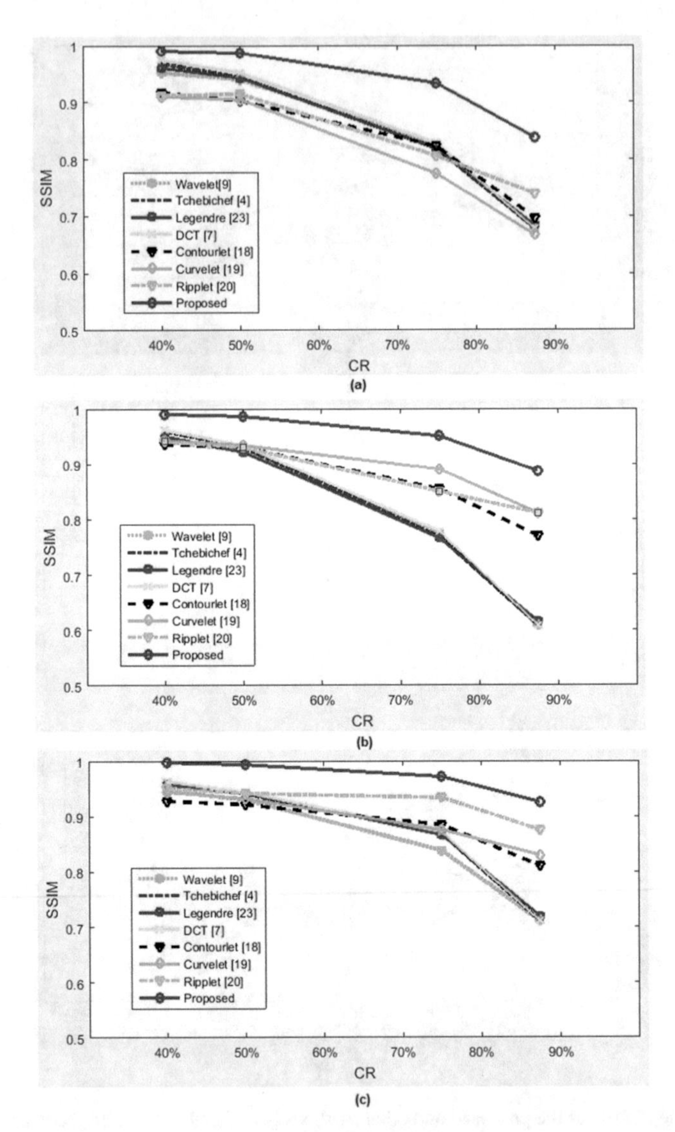

Fig. 7.9 The SSIM for the proposed and other methods [4], [7], [9], [18], [19], [20], and [23]: (**a**) Abdomen_CT.dcm image, (**b**) Leg_MRI.dcm image and (**c**) Chest_Xray.dcm image

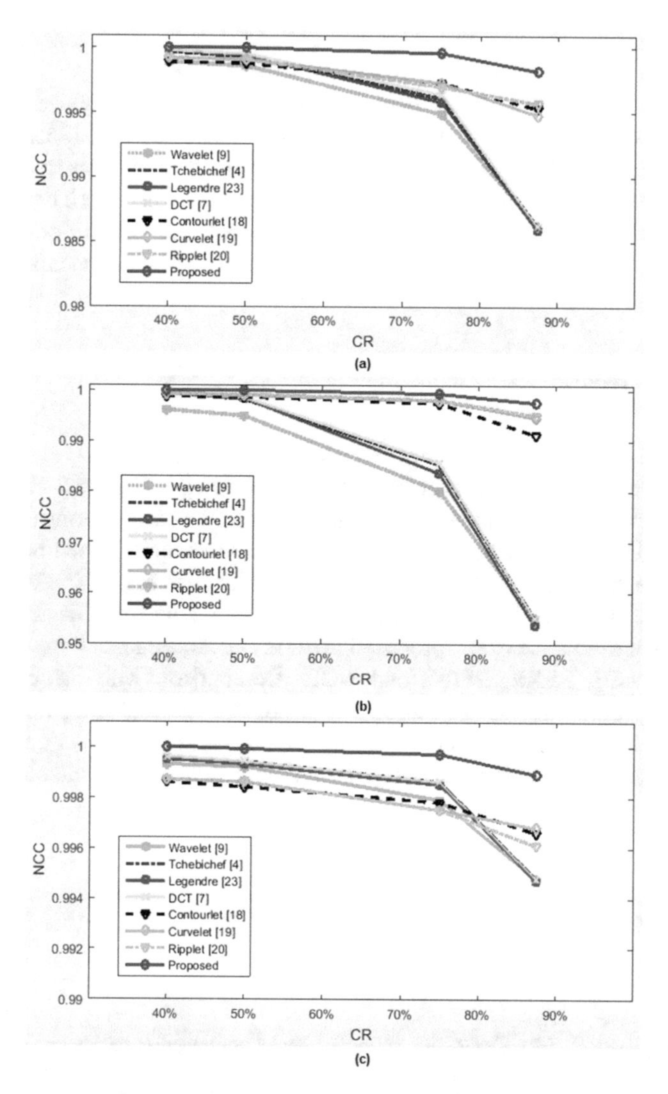

Fig. 7.10 The NCC for the proposed and other methods [4], [7], [9], [18], [19], [20], and [23]: (**a**) Abdomen_CT.dcm image, (**b**) Leg_MRI.dcm image, and (**c**) Chest_Xray.dcm image

7.6 Limitations of the Proposed Algorithm

In comparison with other compression algorithms in terms of implementation time, the proposed algorithm is relatively slow. This slowness is due to the implementation of the WOA algorithm. In WOA, a large number of iterations are elapsed in order to search for the optimum moments for each block. However, the proposed algorithm could be significantly accelerated by utilizing high speed processors or multicore CPUs and GPUs parallel architecture in computer systems.

In the proposed compression algorithm, the image can be divided into blocks of different sizes 8×8, 16×16, or 32×32 which are standard block sizes used in many other techniques. However, increasing block size requires to increase the number of iterations needed to find the optimum coefficients in each bock in WOA, which makes the algorithm to be slower. So, we preferred to use block size 8×8 which leads to efficient compression results and reasonable implementation time.

7.7 Conclusion

In this paper, the WOA algorithm successfully utilized with Legendre moments for optimizing the compression of 2D medical images. Medical images from different medical imaging modalities are used in testing and evaluating the proposed algorithm. The input medical image is divided into nonoverlapped block of size 8×8, where the Legendre moments are computed for each block. The WOA is used to select the optimum Legendre moments by minimizing the objective function, MSE. The performance of the proposed method is evaluated using the quantitative measures, MSE, PSNR, SSIM, and NCC. Comparison with the conventional Legendre method and the well-known methods for medical image compression is performed, the proposed method achieves the highest reconstruction quality for all CRs. The proposed medical compression algorithm is very suitable for recent applications in telemedicine.

References

1. Bronzino, Joseph D.: Biomedical engineering handbook: biomedical engineering fundamentals. CRC press. **2**, 1999.
2. Wootton, Richard, John Craig, and Victor Patterson: Introduction to telemedicine. London: Royal Society of Medicine Press. **206**, 2006.
3. Tom Broens, Rianne Huis in t Veld, Miriam M Vollenbroek, and Bart Nieuwenhuis: Determinants of successful telemedicine implementations: a literature study. Journal of telemedicine and telecare. **6**, 303-309, 2007.
4. Begum, AH Ragamathunisa, D. Manimegalai, and A. Abudhahir: Optimum coefficients of discrete orthogonal Tchebichef moment transform to improve the performance of image compression. Malaysian Journal of Computer Science. **26(1)**, 60-75, 2013.
5. Manjari Singh, Sushil Kumar, Siddharth Singh, and Manish: Various Image Compression Techniques: Lossy and Lossless. International Journal of Computer Applications. **142(6)**, 2016.
6. Wu, Yung-Gi: Medical image compression by sampling DCT coefficients. IEEE Transactions on Information Technology in Biomedicine. **6(1)**, 86-94, 2002.
7. Chen, Yen-Yu: Medical image compression using DCT-based subband decomposition and modified SPIHT data organization. International journal of medical informatics. **76(10)**, 717-725, 2007.
8. Poularikas, Alexander D.: Transforms and applications handbook. CRC Press, 2010.

9. Rashmita Sahoo, Sangita Roy, and Sheli Sinha Chaudhuri: Haar Wavelet Transform image compression using Run Length Encoding. Communications and Image Processing (ICCSP), 2014 International Conference on. IEEE, 2014.
10. Talukder, Kamrul Hasan, and Koichi Harada: Haar wavelet-based approach for image compression and quality assessment of compressed image. IAENG International Journal of Applied Mathematics. **36(1)**, 1-8, 2007.
11. I.UrrizaJ, I.ArtigasL, A.BarraganJ, I.GarciaD, and Navarro: Medical image compression using DCT-based subband decomposition and modified SPIHT data organization. Real-Time Imaging. **7(2)**, 203-217, 2001.
12. Chowdhury, M. Mozammel Hoque, and Amina Khatun: Image compression using discrete wavelet transform. IJCSI International Journal of Computer Science Issues. **9(4)**, 327-330, 2012.
13. Stephane Mallat: A Wavelet Tour of Signal Processing. 2nd edition. Academic Press, New York, 1999.
14. Starck J.L., Candes, E.J., Donoho, D.L.: Curvelets, multiresolution representation, and scaling laws. IEEE Trans. Image Process. **11**, 670-684, 2000.
15. Do, M.N., Vetterli, M.: The contourlet transform: an efficient directional multiresolution image representation. IEEE Trans. Image Process. **14(12)**, 2091-2106, 2005.
16. Jun Xu, Lei Yang and Dapeng Wu: Ripplet: A New Transform for Image Processing. Journal of Visual Communication and Image Representation. **21(7)**, 627-639, 2010.
17. Eslami, R., Radha, H.: Wavelet-based contourlet coding using an SPIHT-like algorithm. In: Proceeding of the conference on information sciences and systems, Princeton. 784-788, 2004.
18. Hashemi-berenjabad, Seyyed Hadi, and Ali Mahloojifar: A New Contourlet-based Compression and Speckle Reduction Method for Medical Ultrasound Images. International Journal of Computer Applications. **82(13)**, 26-32, 2013.
19. Anandan, P., and R. S. Sabeenian: Medical Image Compression Using Wrapping Based Fast Discrete Curvelet Transform and Arithmetic Coding. Circuits and Systems. **7(8)**, 2059-2069, 2016.
20. Juliet Sujitha, Elijah Blessing Rajsingh, and Kirubakaran Ezra: A novel medical image compression using ripplet transform. Journal of Real-Time Image Processing. **11(2)**, 401-412, 2016.
21. Khalid M. Hosny, Asmaa M. Khalid, and Ehab R. Mohamed: Efficient compression of bio-signals by using Tchebichef moments and Artificial Bee Colony. Biocybernetics and Biomedical Engineering. **38**, 385-398, 2018.
22. Lazaridis Pavlos, Bizopoulos, A., Tzekis, P., Zaharis, Z., Debarge, G., and Gallion, P.: Comparative study of DCT and discrete Legendre transform for image compression. X International Conference on Electronics, Telecommunications, Automatics and Informatics (ETAI 2011), At Ohrid, FYR of Macedonia, 2011.
23. K. H. ThungS. C. NgC. L. LimP. Raveendran: A preliminary study of compression efficiency and noise robustness of orthogonal moments on medical X-Ray images.5th Kuala Lumpur International Conference on Biomedical Engineering 2011. Springer Berlin Heidelberg, 2011.
24. Khalid M. Hosny: Exact Legendre moment computation for gray level images. Pattern Recognition. **40(12)**, 3597-3605, 2007.
25. Khalid M. Hosny: Robust template matching using orthogonal Legendre moment invariants. Journal of Computer Science. **6(10)**, 1080-1084, 2010.
26. Srinivasa Rao, Srinivas Kumar, Chandra Mohan: Content Based Image Retrieval Using Exact Legendre Moments and Support Vector Machine. International Journal of Multimedia & Its Applications. **2(2)**, 69-79, 2010.
27. Hussein Mustapha, and Roussos Dimitrakopoulos: HOSIM: A high-order stochastic simulation algorithm for generating three-dimensional complex geological patterns. Computers & Geosciences. **37(9)**, 1242-1253, 2011.
28. Nakib, Schulze, and Petit: Image thresholding framework based on two-dimensional digital fractional integration and Legendre moments. IET Image Processing. **6(8)**,717-727, 2012.

29. Sonia Dahdouha, Elsa D. Angelinia, Gilles Grangeb, and Isabelle Blocha: Segmentation of embryonic and fetal 3D ultrasound images based on pixel intensity distributions and shape priors. Medical Image Analysis. **24(1)**, 255-268, 2015.
30. Khalid M. Hosny, George A. Papakostas, and Dimitris E. Koulouriotis: Accurate reconstruction of noisy medical images using orthogonal moments. Digital Signal Processing (DSP), 2013 18th International Conference on. IEEE, 2013.
31. Khalid M. Hosny: Refined translation and scale Legendre moment invariants. Pattern Recognition Letters. **31(7)**, 533-538, 2010.
32. Mirjalili, Seyedali, and Andrew Lewis: The whale optimization algorithm. Advances in engineering software. **95**, 51-67, 2016.
33. Jeremy A. Goldbogen, Ari S. Friedlaender, John Calambokidis, Megan F. McKenna, Malene Simon, Douglas P. Nowacek: Integrative Approaches to the Study of Baleen Whale Diving Behavior, Feeding Performance, and Foraging Ecology. BioScience. **63**, 90-100, 2013.
34. A. Kaveh and M. Ilchi Ghazaan: Enhanced whale optimization algorithm for sizing optimization of skeletal structures. Mechanics Based Design of Structures and Machines. **45(3)**, 345-362, 2017.
35. A. Kaveh and M. Ilchi Ghazaan: Enhanced whale optimization algorithm for sizing optimization of skeletal structures. Mechanics Based Design of Structures and Machines. **45(3)**, 345-362, 2017.
36. https://www.dicomlibrary.com/.

Chapter 8
Online Variational Learning Using Finite Generalized Inverted Dirichlet Mixture Model with Feature Selection on Medical Data Sets

Meeta Kalra and Nizar Bouguila

Abstract At present, wide applications of clustering algorithms in biomedical research can be witnessed on high-dimensional data. Few examples include gene expression profiling using micro-array data, morphological feature extraction from cellular images, and medical imaging analysis on MRI data. Henceforth, it is of utmost importance to have the right match between the biomedical data and the applied algorithm. For such diverse goals, specialized clustering algorithms are developed in association with the biomedical information, terminologies, and requirements captured in the snapshot of the image. For this, clustering models are developed from the domains of machine learning, data mining, and image processing which focus on extracting features from the data sets and analyzing these models by using diverse approaches.

In this chapter, we propose a statistical framework for online variational learning of finite generalized inverted Dirichlet (GID) mixture model for clustering medical images data by simultaneously using feature selection and image segmentation. The model allows one to adjust the mixture model parameters, number of components and features weights to tackle the challenge of over-fitting. The algorithm in this study has been evaluated on synthetic data as well as three medical applications for brain tumor detection, skin melanoma detection, and computer-aided detection (CAD) of malaria.

M. Kalra (✉) · N. Bouguila
Concordia Institute for Information Systems Engineering, Concordia University, Montreal, QC, Canada
e-mail: m_lra@encs.concordia.ca; nizar.bouguila@concordia.ca

M. Masmoudi et al. (eds.), *Artificial Intelligence and Data Mining in Healthcare*,
https://doi.org/10.1007/978-3-030-45240-7_8

8.1 Introduction

In today's technological era, data is growing enormously irrespective of the domain from which it is evolving. The major challenge is the statistical modeling, processing, and analyzing of these data sets to interpret them without bias. For this reason, there is a constant endeavor in developing machine learning techniques. Medicine is one such area where immensely valuable heterogeneous data is generated. These data have been the foundation of research to evaluate various machine learning algorithms.

Biomedical and anatomical information are majorly easy to obtain as a result of the success achieved in automating image segmentation. Computer image segmentation algorithms for the delineation of anatomical structures and other regions of interest are key components assisting and automating specific radiological tasks. There are different methods employed for medical image segmentation such as classifier, Markov random model, region growing, thresholding method, and clustering methods. These methods are chosen based on different image modalities, image artifacts such as noise, motion, and also specific applications. Among these methods, in our chapter we would be focusing on clustering methods in which the standard methods of K-means clustering [1, 2], fuzzy C, and fuzzy K-C-means clustering [3] have already been in use for various medical practices to help in diagnosis. However in this chapter, we propose a different clustering method with online variational learning of finite generalized inverted Dirichlet (GID) mixture model with feature selection. Finite mixture models have been extensively used in the medical area for various purposes. It has been used for tracking hospital activity and management of healthcare by using length of hospital stay [4], for brain tumor detection, skin melanoma detection [5], and also to propose different image segmentation methods for segmenting different medical images [6].

Clustering is an unsupervised machine learning approach where the aim is to segregate observations with similar traits and assign them into clusters. It is basically dividing the data points into a number of groups such that data points with similar characteristics are grouped in the same cluster as compared to other data points with dissimilar characteristics which are grouped in a different cluster. It is an extensive research topic in which many parametric and non-parametric approaches have been proposed. The implementation of clustering theories and algorithms on biomedical data constitutes an important portion of the rapidly developing field of research. To this end, in this chapter we provide a step-by-step guidance for biomedical researchers to our proposed data pre-processing using feature selection and an algorithm which is specially designed for medical image data sets. Therefore, we propose the implementation of such clustering models in the daily work-flow of medical data analysis [7].

Mixture models are a very popular statistical based data clustering approach which makes inferences via probabilistic assumptions of the data distribution. Thus, it is extremely crucial to choose the most proper distribution that best represents the corresponding components of the mixture accurately when modeling data. One of the widely applied approaches for modeling multidimensional data combinations in several fields is the Gaussian mixture [8–12]. However, real life data come in with many different properties, many of which can be clearly seen as non-Gaussian, such as proportional data, for which Dirichlet family of distributions has been proven to be a more acclaimed choice for cluster analysis [10, 13]. In order to understand data, researchers always try to describe it by its features. Hence, clustering data based on its features along with model selection are one of the vital means of analysis. Indeed, it has been shown that not all the features have the same contribution in the clustering process such as in [14] where the generalized Dirichlet (GD) mixture has been used and factorized into a set of Beta distributions giving good results for proportional data [15] clustering. Though, the Beta distribution support is defined in $[0\ 1]$, so it is not always an appropriate choice to represent positive vectors. Therefore, the GID mixture is a more adequate choice to represent positive vectors since it can be factorized into a set of Beta prime (inverted Beta) distributions whose support is $[0\ \infty]$. For such instances, GID mixture models have proved to be an efficient choice over Gaussian mixture models [10, 16–18].

Generally, there are two challenging problems when dealing with finite mixture models. Firstly, the estimation of the mixture parameters and secondly, the determination of the number of components. Parameters can be estimated by either of the two families of procedures: the Bayesian or the frequentist. In the frequentist approach, maximum likelihood is the most popular. While in the landscape of multidimensional parameter space, maximum likelihood can falsely get held in packsaddle points or regional maxima leading to incorrect estimation of the model complexity. This false solution can frequently lead to an over-fitting of the model and poor parameter estimation for a given model. Therefore, Bayesian framework can address the aforementioned inadequacies. The strength of Bayesian learning lies in considering the parameters as random variables and then describing our knowledge on the probabilistic model by following probability distributions (known as *priors*) of the model parameters. A posterior distribution of each parameter is estimated by considering the prior of that parameter. The main feature of Bayesian approach is that it permits explicit utilization of previous information and is able to provide an understanding of where classical statistics fail. Furthermore, they do not suffer from over-fitting problem. Despite that, learning using fully Bayesian approaches is computationally challenging and time-consuming. This has enforced the consideration of other approximation techniques using Bayesian inference, such as Laplace approximation and Markov chain Monte Carlo (MCMC). In Laplace approach, the approximation for the integrals expression is calculated by utilizing the expansion of Taylor series [16].

The approximation technique recognized as variational Bayes or variational inference is then introduced with the inherited strengths from conventional Bayesian while avoiding its disadvantages [19]. Its main idea is based on using an approximated variant of the true posterior distribution and requires only a modest amount of computational power. It has a tractable inference process that consists of estimating a lower bound for the likelihood of observed data with a marginalization performed over unobserved variables [20]. Variational algorithms are warranted to supply a lower limit on the error of approximation [21, 22]. Variational learning is considered to be more fitting while we are dealing with non-Gaussian data and making it possible to suit a big class of learning models and then to discover data with real-world complexity [23, 24]. Variational approximation is set on analytical approaches to model posterior distributions and also provides good generalization performance as well as computational tractability in various applications of finite mixture models. This development has also increased the power and flexibility of finite mixture models by permitting full inference about all the involved parameters and then allows simultaneous model selection and parameters estimation. When comparing with batch variational, algorithms for online mixture learning are more effective when handling large and continuous data. In recent times, models based on online Gaussian mixture and its expansion have been evaluated [25]. However, as we mentioned before the Gaussian supposition is infrequently met and is not realistic in several applications of real life. In actuality, the new works have demonstrated that other models may give better modeling capacities for non-Gaussian data. For example, Dirichlet mixtures can be a better substitution in many applications particularly those including proportional data [26].

Thus, in this chapter we propose an online variational learning framework which has been used in different fields of image processing and computer vision. An advantage of GID is the fact that it has a more general covariance structure than the inverted Dirichlet whose covariance is strictly positive. We structure lower bound that could be traced for estimating marginal probability by utilizing approximated divisions to exchange distributions of the intractable parameter. This algorithm is much more effective and has fast convergence. Moreover, compared with conventional techniques, in which the selection of model is resolved based upon cross-validation, our algorithm estimates the parameters of model and limits the number of components concomitantly. It is worth noticing that the challenge of designing an online learning algorithm of GID with feature selection is to consider both model complexity (i.e., number of clusters) and feature relevancy where some irrelevant features may become relevant and vice versa when each time new vectors are added [27]. Some previous works have considered the problem of streaming feature selection (for instance, [28, 29]) where the set of observations are supposed to be fixed, but the set of features is considered dynamic and not known in advance. However, our research efforts focus on an assumption where we assume that the set of features is fixed, while the set of observations is dynamic.

The rest of this chapter is organized as follows: Sect. 8.2 discusses clustering applications in healthcare, Sect. 8.3 introduces the statistical model of finite GID with feature selection concisely. In Sect. 8.4, we describe our online variational Bayes algorithm for the proposed model learning. Section 8.5 presents our proposed algorithm preciseness and effectiveness on sets of synthetic data as well as on biomedical applications for brain tumor detection using MRI image segmentation, skin melanoma detection using skin image segmentation, and CAD malaria detection using feature selection. Lastly, we conclude the chapter in Sect. 8.6.

8.2 Clustering Applications in Healthcare

Cluster analysis (CA) helps revealing structures that are hidden by gathering entities or objects (e.g., products, locations, individuals) with similar features into identical groups while making as great as possible heterogeneity among groups [30, 31]. Objects or entities of interest are gathered together and grounded in properties, which make them comparable together with the last purpose being to classify these objects or entities by means of clustering them into similar groups and to take them apart from various groups. In terms of a theoretical concept, CA is designed to recognize cluster solutions that are approximately identical inside every group, resulting in clusters that have a big intra-class resemblance while maximizing dissimilarity through the groups [32]. In recent years, the exponential growth of data necessitates in developing modern methods which can manage large numbers of saved information and data and this is especially correct in the health-care domain. Thus, CA has widely been used in different applications involving finding a correct typology, the expectation based upon groups, generation of hypothesis, exploration of data, and reduction of data or gathering identical entities within comparable classes, accordingly organizing huge amounts of information and allowing labels to ease communication [30, 33, 34] such as identifying patients of based on symptoms clusters [35]; by finding genes group which have identical biological functions [36]; or determining groups of patients [33, 34].

Few examples where CA has been applied in healthcare domain are as follows. Many researches have been performed to distinguish brain tumor cells with other normal brain cells by using the methodology of brain tumor segmentation with Fuzzy C-Means clustering as a hybrid approach [37]. In [38] brain MRI segmentation has been done using clustering and features selection approach. Yao et al. [39] have used texture features, support vector machine model, and wavelet transform to achieve the excellent classification of dynamic MRI images and to deal with non-linear real data. The finite mixture model has also been used in gynecology and obstetrics research to analyze the pregnancy costs [40]. In order to analyze patient health care costs and utilizations, the choice of a model is critical for a clear estimation of quantities like incremental costs.

In this chapter, we have used our proposed clustering algorithm on different medical data set to evaluate its performance in diagnosing specific diseases. The algorithm worked very well in detecting the brain tumor, skin melanoma, and CAD of malaria.

8.3 Model Specification

In the previous section, we discussed the importance of data clustering and saw the huge importance it has in health care applications. The below section gives an overview of the finite generalized inverted Dirichlet (GID) mixture model with the unsupervised feature selection approach previously proposed in [14] and defines an online variational framework for it, which allows estimating the parameters and the number of components of the mixture model automatically and simultaneously.

8.3.1 Finite Generalized Inverted Dirichlet Mixture Model with Feature Selection

The most significant reason to consider generalized inverted Dirichlet distribution as a standard one in our mixture model is its ability to generate models specified to positive vectors and its more general covariance structure. The GID has several interesting mathematical properties which allow, for instance, the representation of GID samples in a transformed space in which features are independent and follow inverted Beta distributions [41]. We consider a set $\mathcal{Y}$ of N D-dimensional positive vectors, such that $\mathcal{Y} = (Y_1, Y_2, ..., Y_N)$ and M indicates the number of various clusters [42]. We suppose that $\mathcal{Y}$ is managed by a mixture of GID distributions $p(\mathcal{Y}_i \mid \boldsymbol{\pi}, \boldsymbol{\alpha}, \boldsymbol{\beta},)$ [17] as

$$p(\mathcal{Y}_i \mid \boldsymbol{\pi}, \boldsymbol{\alpha}, \boldsymbol{\beta}) = \sum_{j=1}^{M} \pi_j \prod_{l=1}^{D} \frac{\Gamma(\alpha_{jd} + \beta_{jd})}{\Gamma(\alpha_{jd})\Gamma(\beta_{jd})} \frac{\mathcal{Y}_{id}^{\alpha_{jd}-1}}{\left(1 + \sum_{l=1}^{d} \mathcal{Y}_{il}\right)^{\gamma_{jd}}} \tag{8.1}$$

where $\boldsymbol{\alpha} = \{\alpha_1, \alpha_2, \ldots \alpha_M\}$, with $\boldsymbol{\alpha}_j = \{\alpha_{j1}, \alpha_{j2}, \ldots \alpha_{jD}\}$, $j = 1, \ldots, M$ and $\boldsymbol{\beta} = \{\beta_1, \beta_2, \ldots \beta_M\}$, with $\boldsymbol{\beta}_j = \{\beta_{j1}, \beta_{j2}, \ldots i\beta_{jD}\}$, $j = 1, \ldots, M$. $\boldsymbol{\pi} = \{\pi_1, \pi_2, \ldots \pi_M\}$, are the mixing weights, such that $\sum_{j=1}^{M} \boldsymbol{\pi}_\mathbf{j} = 1$. We define $\boldsymbol{\gamma}_\mathbf{jd}$ such that $\boldsymbol{\gamma}_\mathbf{jd} = \boldsymbol{\beta}_\mathbf{jd} + \boldsymbol{\alpha}_\mathbf{jd} - \boldsymbol{\beta}_{\mathbf{j}(\mathbf{d+1})}$. The GID posterior probability can be factorized as follows [17]:

$$p(j \mid \mathcal{Y}_i, \boldsymbol{\pi}, \boldsymbol{\alpha}, \boldsymbol{\beta}) \propto \pi_j \prod_{l=1}^{D} p_{iBeta}(\mathcal{X}_{il} \mid \boldsymbol{\alpha}_\mathbf{jl}, \boldsymbol{\beta}_\mathbf{jl}) \tag{8.2}$$

where we have set $\mathcal{X}_{il} = \mathcal{Y}_{il}$ and $\mathcal{X}_{il} = \frac{\mathcal{Y}_{il}}{1+\sum_{k=1}^{l=1} \mathcal{Y}_{ik}}$ for $\mathbf{l} > 1$. $p_{iBeta}(\mathcal{X}_{il} \mid \boldsymbol{\alpha}_\mathbf{jl}, \boldsymbol{\beta}_\mathbf{jl})$ is an inverted Beta distribution with parameters $\boldsymbol{\alpha}_\mathbf{jl}$ and $\boldsymbol{\beta}_\mathbf{jl}$ as below:

$$p_{iBeta}(\mathcal{X}_{il} \mid \boldsymbol{\alpha}_\mathbf{jl}, \boldsymbol{\beta}_\mathbf{jl}) = \frac{\Gamma(\alpha_{jd} + \beta_{jd})}{\Gamma(\alpha_{jd})\Gamma(\beta jd)} \mathcal{X}_{il}^{\alpha_{jl}-1} (1 + \mathcal{X}_{il})^{-(\boldsymbol{\alpha}_\mathbf{jl} + \boldsymbol{\beta}_\mathbf{jl})} \tag{8.3}$$

Let $\vec{\mathcal{Z}}_i = (\mathbf{Z_{i1}}, \ldots, \mathbf{Z_{iM}})$ be a binary latent variable assigned to each observation $\vec{\mathbf{X}}_i$. The values of $\mathcal{Z}_i$ satisfy $\mathcal{Z}_{ij} \in \{0, 1\}$, $\sum_{j=1}^{M} \mathcal{Z}_{ij} = 1$, $\mathcal{Z}_{ij} = 1$ if $\vec{\mathbf{X}}_i$ belongs to component j and equal to 0, otherwise. The conditional distribution of latent variables $\mathcal{Z} = (\vec{\mathbf{Z}}_1, \ldots, \vec{\mathbf{Z}}_N)$ given the mixing coefficients $\vec{\pi}$ can be written as

$$p(\mathcal{Z} \mid \vec{\pi}) = \prod_{i=1}^{N} \prod_{j=1}^{M} \boldsymbol{\pi}_{\mathbf{j}}^{Z_{ij}} \tag{8.4}$$

Thus, given the latent variables and the component parameters set we are able to write the conditional distribution of the data set $\mathcal{X} = (\vec{\mathbf{X}}_1, \ldots, \vec{\mathbf{X}}_N)$ as:

$$p(\mathcal{X} \mid \mathcal{Z}, \vec{\alpha}, \vec{\beta}) = \prod_{i=1}^{N} \prod_{j=1}^{M} \left(\prod_{l=1}^{D} iBeta(\mathcal{X}_{il} \mid \boldsymbol{\alpha}_{\mathbf{jl}}, \boldsymbol{\beta}_{\mathbf{jl}}) \right)^{Z_{ij}} \tag{8.5}$$

Feature selection is an important aspect when data is multidimensional and some features could be noisy, which can impact the algorithm performance as well as the clustering process. These features can thus be considered irrelevant since they do not have any discriminatory impact on the clustering. As so, to integrate feature selection with finite GID mixture model in Eq. (8.2) and to take into consideration the fact that the features $\mathbf{X}_{il}$ are mostly not equally important for the clustering task, the following approximation for the $\mathcal{X}_{il}$ distribution has been suggested:

$$p(\mathcal{X}_{il} \mid \mathcal{W}_{ikl}, \boldsymbol{\phi}_{\mathbf{il}}, \boldsymbol{\alpha}_{\mathbf{jl}}, \boldsymbol{\beta}_{\mathbf{jl}}, \boldsymbol{\lambda}_{\mathbf{kl}}, \boldsymbol{\tau}_{\mathbf{kl}}) \simeq iBeta(\mathcal{X}_{il} \mid \boldsymbol{\alpha}_{jl}, \boldsymbol{\beta}_{jl})^{\phi_{il}}) \left(\prod_{K=1}^{K} iBeta(\mathcal{X}_{il} \mid \boldsymbol{\lambda}_{\mathbf{kl}}, \boldsymbol{\tau}_{\mathbf{kl}})^{W_{ikl}} \right)^{1-\phi il} \tag{8.6}$$

where $\boldsymbol{\phi}_{\mathbf{il}}$ is a binary latent variable, such that $\boldsymbol{\phi}_{\mathbf{il}} = 1$ indicates that l is relevant feature and follows an inverted Beta distribution iBeta $(\mathcal{X}_{il}|\alpha_{jl}, \beta_{jl})$. However, $\boldsymbol{\phi}_{\mathbf{il}} = 0$ represents that feature l is irrelevant and supposed to follow a finite mixture of inverted beta distributions independent from the class labels such as:

$$p(\mathcal{X}_{il}) = \sum_{K=1}^{K} \eta_{kl} iBeta(\mathcal{X}_{il} \mid \boldsymbol{\lambda}_{\mathbf{kl}}, \boldsymbol{\tau}_{\mathbf{kl}}) \tag{8.7}$$

where n_{kl} denotes a mixing probability and implies the prior probability that $\mathcal{X}_{il}$ is generated from the kth component of the finite inverted beta mixture representing irrelevant features, and $\sum_{K=1}^{K} \eta_{kl} = 1$.

In Eq. (8.6), $\mathcal{W}_{ikl}$ is a binary latent variable such that $\mathcal{W}_{ikl} = 1$ only if $\mathcal{X}_{il}$ comes from the kth component of the finite inverted beta mixture for the irrelevant features. The conditional distribution of the latent variables $\mathcal{W} = (\vec{W}_1, \ldots, \vec{W}_N)$ with $\vec{\mathcal{W}}_i = (\vec{W}_{i1}, \ldots, \vec{W}_{iK})$ and $\vec{\mathcal{W}}_{ik} = (\vec{W}_{ik1}, \ldots, \vec{W}_{ikD})$ given the mixing coefficients $\vec{\eta}$ can be written as

$$p(\mathcal{W} \mid \vec{\eta}) = \prod_{i=1}^{N} \prod_{K=1}^{K} \prod_{L=1}^{D} \boldsymbol{\eta_{kl}}^{W_{ikl}} \tag{8.8}$$

where $\vec{\eta} = (\vec{\eta}_1, \ldots, \vec{\eta}_K)$ with element $\vec{\eta}_k = (\vec{\eta}_{k1}, \ldots, \vec{\eta}_{kD})$. The conditional distribution of the feature relevancy indicator variable $\vec{\phi} = (\vec{\phi}_1, \ldots, \vec{\phi}_N)$ with elements $(\vec{\phi}_{i1}, \ldots, \vec{\phi}_{iD})$, given $\vec{\epsilon}$, is defined as

$$p(\vec{\phi} \mid \vec{\epsilon}) = \prod_{i=1}^{N} \prod_{l=1}^{D} \boldsymbol{\epsilon_{l_1}}^{\phi_{il}} \boldsymbol{\epsilon_{l_2}}^{1-\phi_{il}} \tag{8.9}$$

where ϕ is a Bernoulli variable such that $p(\phi_{il} = 1) = \epsilon_{l_1}$ and $p(\phi_{il} = 0) = \epsilon_{l_2}$. The vector $\vec{\epsilon} = (\vec{\epsilon}_1, \ldots, \vec{\epsilon}_D)$ represents the probabilities of the relevant features called feature saliencies such that $\vec{\epsilon}_l = (\epsilon_{l_1}, \epsilon_{l_2})$ and $(\epsilon_{l_1} + \epsilon_{l_2}) = 1$. Therefore, the likelihood of the observed data set $\mathcal{X}$ following the finite GID mixture model with feature selection is given as follows:

$$p(\mathcal{X} \mid \mathcal{Z}, \mathcal{W}, \vec{\phi}, \vec{\alpha}, \vec{\beta}, \vec{\lambda}, \vec{\tau}) = \prod_{i=1}^{N} \prod_{j=1}^{M} \Bigg[\prod_{l=1}^{D} iBeta(\mathbf{X_{il}} \mid \boldsymbol{\alpha_{il}}, \boldsymbol{\beta_{jl}})^{\phi_{il}} \times \Bigg(\prod_{K=1}^{K} iBeta(\mathbf{X_{il}} \mid \boldsymbol{\lambda_{kl}}, \boldsymbol{\tau_{kl}})^{W_{ikl}} \Bigg]^{1-\phi_{il}} \Bigg)^{Z_{ij}} \tag{8.10}$$

The detailed description of this unsupervised feature selection model is given in [14].

8.3.2 *Prior Specifications*

The setting up of prior distributions is a very crucial step in variational learning. Hence, we have to place priors over ($\vec{\alpha}$), ($\vec{\beta}$), ($\vec{\lambda}$), and ($\vec{\tau}$). The consideration of conjugate priors is the key factor which majorly simplifies variational inference method. In our case, we consider the gamma distribution to approximate a Beta distribution conjugate prior as suggested in [43] which gives the following priors:

$$p(\vec{\alpha}) = \mathcal{G}(\vec{\alpha} \mid \vec{u}, \vec{v}) = \prod_{j=1}^{M} \prod_{l=1}^{D} \frac{v_{jl}^{u_{jl}}}{\Gamma(u_{jl})} \alpha_{jl}^{u_{jl}-1} e^{-v_{jl}\alpha_{jl}} \tag{8.11}$$

$$p(\vec{\beta}) = \mathcal{G}(\vec{\beta} \mid \vec{p}, \vec{q}) = \prod_{j=1}^{M} \prod_{l=1}^{D} \frac{q_{jl}^{p_{jl}}}{\Gamma(p_{jl})} \beta_{jl}^{p_{jl}-1} e^{-q_{jl}\beta_{jl}} \tag{8.12}$$

$$p(\vec{\lambda}) = \mathcal{G}(\vec{\lambda} \mid \vec{g}, \vec{h}) = \prod_{K=1}^{M} \prod_{l=1}^{D} \frac{h_{kl}^{g_{kl}}}{\Gamma(g_{kl})} \lambda_{kl}^{g_{kl}-1} e^{-h_{kl}\lambda_{kl}} \tag{8.13}$$

$$p(\vec{\tau}) = \mathcal{G}(\vec{\tau} \mid \vec{s}, \vec{t}) = \prod_{K=1}^{M} \prod_{l=1}^{D} \frac{t_{sl}^{s_{kl}}}{\Gamma(s_{kl})} \tau_{kl}^{s_{kl}-1} e^{-t_{kl}\tau_{kl}} \tag{8.14}$$

where all the hyper-parameters $\vec{u} = \{u_{jl}\}$, $\vec{v} = \{v_{jl}\}$, $\vec{p} = \{p_{jl}\}$, $\vec{q} = \{q_{jl}\}$, $\vec{g} = \{g_{kl}\}$, $\vec{h} = \{h_{kl}\}$, $\vec{s} = \{s_{kl}\}$, and $\vec{t} = \{t_{kl}\}$ of the above conjugate priors are positive. We do not consider $\vec{\pi}$, $\vec{\eta}$, and $\vec{\epsilon}$ as random variables in our model so no priors are considered for them. The joint distribution of all the random variables for GID mixture model with feature selection is given by

$$\begin{aligned} p(\mathcal{X} \mid \mathcal{Z}, \mathcal{W}, \vec{\phi}, \vec{\alpha}, \vec{\beta}, \vec{\lambda}, \vec{\tau}) &= p(\mathcal{X} \mid \mathcal{Z}, \mathcal{W}, \vec{\phi}, \vec{\alpha}, \vec{\beta}, \vec{\lambda}, \vec{\tau}) \\ &\times p(\mathcal{Z} \mid \vec{\pi}) p(\mathcal{W} \mid \vec{\eta}) p(\vec{\phi} \mid \vec{\epsilon}) p(\vec{\alpha}) p(\vec{\beta}) p(\vec{\lambda}) p(\vec{\tau}) \end{aligned} \tag{8.15}$$

8.4 Online Variational Learning for Finite Generalized Inverted Dirichlet Mixture Mode with Feature Selection

Variational procedures are very common and have been extensively utilized in the past to find approximations which are tractable for posterior distributions for a variety of statistical models [44]. One of the most integral parts of designing finite mixture models is parameter estimation and to select the number of components correctly. In this section, we adopt an online variational framework of the finite GID mixture model for parameter estimation and model selection. The online variational concept is taken into account for the dynamic nature of real-world data sets where the observations are sequential. Figure 8.1 represents the graphical representation of our model.

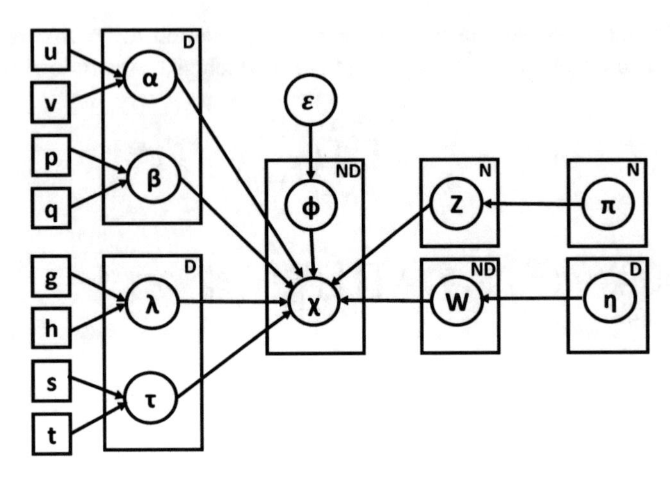

Fig. 8.1 Graphical representation of finite GID mixture model with feature selection. The circles represent the random variables and model parameters. Numbers in the upper right corners of the plates indicate the number of repetitions

8.4.1 Variational Inference

The goal of variational inference method is to find a probability distribution $Q(\Lambda)$ which approximates the true posterior distribution $p(\Lambda \mid \mathcal{X}, \gamma)$. We achieve this by maximizing the lower bound $\mathcal{L}$ on the evidence of model $p(\mathcal{X}|\gamma)$. This evidence of lower bound $\mathcal{L}$ is taken by applying Jensen's inequality on $p(\mathcal{X}|\gamma)$ [15] as:

$$\begin{aligned} \ln P(\mathcal{X} \mid \gamma) &= \ln \int p(\mathcal{X} \mid \Lambda, \gamma) d\Lambda = \ln \int Q(\Lambda) \left(\frac{p(\mathcal{X} \mid \Lambda, \gamma)}{Q(\Lambda)} \right) d\Lambda \\ &\geq \ln \int Q(\Lambda) \left(\frac{p(\mathcal{X} \mid \Lambda, \gamma)}{Q(\Lambda)} \right) d\Lambda = \mathcal{L}(Q) \end{aligned} \tag{8.16}$$

In theory, the lower bound $\mathcal{L}(Q)$ is maximized when $Q(\lambda) = p(\Lambda|\mathcal{X}, \Gamma)$. However, the actual posterior distribution is usually arithmetically intractable and cannot be directly utilized for variational inference. Hence, we use a factorization hypothesis to limit the form of $Q(\Lambda)$ in our work, such that $Q(\Lambda) = Q(\mathcal{Z})Q(\mathcal{W})Q(\vec{\phi})Q(\vec{\alpha})Q(\vec{\beta})Q(\vec{\lambda})Q(\vec{\tau})$. This hypothesis is commonly known as mean field approximation that comes out of statistical mechanics [45] and has been extensively utilized in the past for many applications (for example, [46]). The general optimal solution for improving each variational factor $Q_d(\Lambda_d)$ is given by [47]

$$Q_d(\Lambda_d) = \frac{exp\langle \ln\ p(\mathcal{X}, \Lambda)\rangle_{\neq d}}{\int exp\langle \ln\ p(\mathcal{X}, \Lambda)\rangle_{\neq d} d\Lambda} \tag{8.17}$$

where $\langle . \rangle_{\neq d}$ denotes the expectation with respect to all distributions $Q_j(\Lambda_j)$ except for $j = d$. However, this method cannot be suitable and adaptable for online variational learning. Therefore in our work, we use an alternative approach which deals with variational inference as a gradient method for deriving the optimization solutions [48]. However, both procedures have equivalent consequences for variational inference and the gradient method is more preferable in case of online variational inference for data coming sequentially.

The core idea is that as the model has conjugate priors, the functional form of the factors in the variational posterior distribution is known. According to this, by using general parametric forms on these distributions, the lower bound can be viewed as a function of the parameters of the distributions. We maximize the lower bound with respect to these parameters in order to obtain the optimization of variational factors. In our algorithm, the functional form of each factor is identical to its conjugate prior distribution, specifically discrete for $\mathcal{Z}$ and $\mathcal{W}$, Bernoulli for $\vec{\phi}$, and gamma for $\vec{\alpha}, \vec{\beta}, \vec{\Lambda}$, and $\vec{\tau}$. Thus, the parametric forms of these variational posterior distributions could be defined as following:

$$Q(\mathcal{Z}) = \prod_{i=1}^{N}\prod_{j=1}^{M} r_{ij}^{Z_{ij}}, \quad Q(\mathcal{W}) = \prod_{i=1}^{N}\prod_{K=1}^{K}\prod_{L=1}^{D} \mathbf{m_{kl}}^{W_{ikl}} \tag{8.18}$$

$$Q(\vec{\phi}) = \prod_{j=1}^{N}\prod_{l=1}^{D} f_{il}^{\phi_{il}} (1 - f_{il})^{1-\phi_{il}} \tag{8.19}$$

$$Q(\vec{\alpha}) = \prod_{j=1}^{M}\prod_{l=1}^{D} \mathcal{G}(\alpha_{jl}|u_{jl}^*, v_{jl}^*), \quad Q(\vec{\beta}) = \prod_{j=1}^{M}\prod_{l=1}^{D} \mathcal{G}(\beta_{jl}|p_{jl}^*, q_{jl}^*) \tag{8.20}$$

$$Q(\vec{\lambda}) = \prod_{k=1}^{M}\prod_{l=1}^{D} \mathcal{G}(\lambda_{kl}|g_{kl}^*, h_{kl}^*), \quad Q(\vec{\gamma}) = \prod_{k=1}^{M}\prod_{l=1}^{D} \mathcal{G}(\gamma_{kl}|s_{kl}^*, t_{kl}^*) \tag{8.21}$$

We can obtain the parameterized lower bound $\mathcal{L}(Q)$ by substituting Eqs. (8.18)–(8.21) into (8.16) as below:

$$\begin{aligned}\mathcal{L}(Q) &= \sum_{\theta} \int Q(\Theta, \Omega) \ln \left(\frac{p(\mathcal{X}, \Theta, \Omega \mid \gamma)}{Q(\Theta, \Omega)} \right) d\Omega \\ &= \left\langle \ln p(\mathcal{X}, \Theta, \Omega \mid \gamma) - \ln Q(\Theta, \Omega) \right\rangle \end{aligned} \tag{8.22}$$

The detailed solution of the above equation is explained [15]. Then, the variational parameters r_{ij}, f_{il}, and m_{ikl} can be calculated by maximizing $\mathcal{L}(Q)$ with respect to these parameters, respectively, where

$$r_{ij} = \frac{\tilde{r}_{ij}}{\sum_{j=1}^{M} \tilde{r}_{ij}}, \quad f_{il} = \frac{\tilde{f}_{il}}{\tilde{f}_{il} + \hat{f}_{il}}, \quad m_{ikl} = \frac{\tilde{m}_{ikl}}{\sum_{k=1}^{k} \tilde{m}_{ikl}} \tag{8.23}$$

with

$$\tilde{r}_{ij} = exp\Bigg\{ \ln \pi_j + \sum_{l=1}^{D} \Bigg\{ f_{il}\big[\tilde{R}_{jl} + (\overline{\alpha}_{jl} - 1)\ln \mathcal{X}_{il} - (\overline{\alpha}_{jl} + \overline{\beta}_{jl})\ln(1 + \mathcal{X}_{il})\big]$$
$$+ (1 - f_{il}) \sum_{k=1}^{K} m_{ikl}\big[\tilde{F}_{kl} + (\overline{\lambda}_{kl} - 1)\ln \mathcal{X}_{il} - (\overline{\lambda}_{kl} + \overline{\tau}_{kl})\ln(1 + \mathcal{X}_{il})\big]\Bigg\}\Bigg\} \tag{8.24}$$

$$\tilde{m}_{ikl} = exp\Big\{ \ln \eta_{kl} + (1 - f_{il})\big[\tilde{F}_{kl} + (\overline{\lambda}_{kl} - 1)\ln \mathcal{X}_{il} - (\overline{\lambda}_{kl} + \overline{\tau}_{kl})\ln(1 + \mathcal{X}_{il})\big]\Big\} \tag{8.25}$$

$$\tilde{f}_{ij} = exp\Big\{ \ln \epsilon_{l_1} + \sum_{j=1}^{M} r_{ij}\big[\tilde{R}_{jl} + (\overline{\alpha}_{jl} - 1)\ln \mathcal{X}_{il} - (\overline{\alpha}_{jl} + \overline{\beta}_{jl})\ln(1 + \mathcal{X}_{il})\big]\Big\} \tag{8.26}$$

$$\hat{f}_{il} = exp\Bigg\{ \ln \epsilon_{l_2} + \Bigg\{ \sum_{K=1}^{K} m_{ikl}\big[\tilde{F}_{kl} + (\overline{\lambda}_{kl} - 1)\ln \mathcal{X}_{il} - (\overline{\lambda}_{k1} + \overline{\tau}_{k1})\ln(1 + \mathcal{X}_{il})\big]\Bigg\}\Bigg\} \tag{8.27}$$

$$\begin{aligned}
\tilde{R} = \ln & \frac{\Gamma(\tilde{\alpha} + \tilde{\beta})}{\Gamma(\overline{\beta})\Gamma(\overline{\alpha})} \\
& + \tilde{\alpha}\Big[\psi(\overline{\alpha} + \tilde{\beta}) - \psi(\overline{\alpha})\Big]\Big[\langle \ln \alpha \rangle - \ln \overline{\alpha}\Big] \\
& + \tilde{\beta}\Big[\psi(\overline{\beta} + \tilde{\alpha}) - \psi(\overline{\beta})\Big]\Big[\langle \ln \beta \rangle - \ln \overline{\beta}\Big] \\
& + 0.5\tilde{\alpha}^2\Big[\psi(\overline{\alpha} + \tilde{\beta}) - \psi(\overline{\alpha})\Big]\Big[\langle \ln \alpha \rangle - \ln \overline{\alpha}\Big]^2 \\
& + 0.5\tilde{\beta}^2\Big[\psi(\overline{\beta} + \tilde{\alpha}) - \psi(\overline{\beta})\Big]\Big[\langle \ln \beta \rangle - \ln \overline{\beta}\Big]^2 \\
& + \tilde{\alpha}\tilde{\beta}\psi(\overline{\alpha} + \tilde{\beta})\Big[\langle \ln \beta \rangle - \ln \overline{\beta}\Big]\Big[\langle \ln \alpha \rangle - \ln \overline{\alpha}\Big]
\end{aligned} \tag{8.28}$$

$$
\begin{aligned}
\tilde{F} = & \ln \frac{\Gamma(\tilde{\lambda}+\tilde{\tau})}{\Gamma(\overline{\tau})\Gamma(\overline{\lambda})} \\
& + \tilde{\lambda}\Big[\psi(\overline{\lambda}+\tilde{\tau}) - \psi(\overline{\lambda})\Big]\Big[\big\langle \ln\lambda\big\rangle - \ln\overline{\lambda}\Big] \\
& + \tilde{\tau}\Big[\psi(\overline{\tau}+\tilde{\lambda}) - \psi(\overline{\tau})\Big]\Big[\big\langle \ln\tau\big\rangle - \ln\overline{\tau}\Big] \\
& + 0.5\tilde{\lambda}^2\Big[\psi'(\overline{\lambda}+\tilde{\tau}) - \psi'(\overline{\lambda})\Big]\Big[\big\langle \ln\lambda\big\rangle - \ln\overline{\lambda}\Big]^2 \\
& + 0.5\tilde{\tau}^2\Big[\psi'(\overline{\tau}+\tilde{\lambda}) - \psi'(\overline{\tau})\Big]\Big[\big\langle \ln\tau\big\rangle - \ln\overline{\tau}\Big]^2 \\
& + \tilde{\lambda}\tilde{\tau}\psi'(\overline{\lambda}+\tilde{\tau})\Big[\big\langle \ln\lambda\big\rangle - \ln\overline{\lambda}\Big]\Big[\big\langle \ln\tau\big\rangle - \ln\overline{\tau}\Big]
\end{aligned}
\tag{8.29}
$$

where $\psi(.)$ is the digamma function that is defined as $\psi(\alpha) = \dfrac{d \ln\Gamma(\alpha)}{d(\alpha)}$.

Similarly, we can obtain the update equations of the hyper-parameters of variational factors α, β, γ, and τ (the details are shown in Appendix). Finally, the mixing coefficients π_{ij}, η_{kl} and the feature salencies ϵ_{l_1} can be calculated as:

$$
\pi_j = \frac{1}{N}\sum_{i=1}^{N} r_{ij}, \quad \eta_{kl} = \frac{1}{N}\sum_{i=1}^{N} m_{ikl}, \quad \epsilon_{l_1} = \frac{1}{N}\sum_{i=1}^{N} f_{il} \tag{8.30}
$$

8.4.2 Online Variational Inference

In this subsection, we propose an online variational learning framework with unsupervised feature selection for finite GID mixture model for sequential data. The proposed algorithm approach of online learning is based upon the variational technique developed in [48] which we consider in our work. Since the data are majorly continuously arriving in an online fashion over time, we need to estimate the variational lower bound related to a fixed quantity of data (N in our case). Firstly, we define the estimated value of the model evidence logarithm $\ln P(\mathcal{X})$ for this finite size of data as below

$$
\big\langle \ln p(\mathcal{X})\big\rangle_{\varphi} = \int \phi(\mathcal{X}) \ln\Big(\int p(\mathcal{X}|\Lambda)p(\Lambda)d(\Lambda)\Big)d\mathcal{X} \tag{8.31}
$$

where $\varphi(X)$ represents the unknown probability distribution for the data observed. Thus, the corresponding expected variational lower bound can be computed using [25]:

$$\begin{aligned}\langle \mathcal{L}(Q)\rangle_{\varphi} &= \left\langle \sum_{\Theta} \int Q(\Theta) Q(\Omega) \ln \left[\frac{\mathcal{P}(\mathcal{X}, \Theta|\Omega)\mathcal{P}(\Omega)}{Q(\Omega)Q(\Theta)}\right] d\Omega \right\rangle_{\varphi} \\ &= N \int Q(\Omega) d\Omega \left\langle \sum_{\theta} Q(\Theta) \ln \left[\frac{\mathcal{P}(\vec{X}, \theta|\Omega)}{Q(\theta)}\right\rangle\right]_{\varphi} \\ &+ \int Q(\Omega) \ln \left[\frac{\mathcal{P}(\Omega)}{Q(\Omega)}\right] d\Omega \end{aligned} \tag{8.32}$$

where $\Theta = (\theta_1, ..., \theta_N)$ with $\theta_i = \{\vec{\mathcal{Z}}_i, \vec{\mathcal{W}}_i, \vec{\phi}_i\}$. Let ι denote the actual amount of observed data; thus, the current lower bound for the observed data can be estimated by

$$\begin{aligned}\mathcal{L}^{(t)}(Q) =& \frac{N}{\iota} \sum_{i=1}^{\iota} \int Q(\iota) \mathrm{d}\iota \sum_{\boldsymbol{\theta}} Q(\boldsymbol{\theta}_i) \ln \left[\frac{\mathcal{P}(\mathcal{X}_i, \boldsymbol{\theta}_i|\boldsymbol{\iota})}{Q(\boldsymbol{\theta}_i)}\right] \\ &+ \int Q(\iota) \ln \left[\frac{\mathcal{P}(\iota)}{Q(\iota)}\right] \mathrm{d}\iota \end{aligned} \tag{8.33}$$

It is worth noticing that ι raises by the time while N is fixed. This is because, as debated in [48], the objective function of our online algorithm is the expected log evidence for a fixed quantity of data as shown in Eq. (8.31). Our proposed online algorithm computes the same amount of data even if the amount of observed data raises. In our case, the observed data are used for improving the quality of estimation of the expected variational lower bound in Eq. (8.16) that approximates the expected log evidence.

The core idea of the online variational algorithm is to maximize the present variational lower bound successively in Eq. (8.33). Suppose that we have already observed the data set $\mathcal{X}^{(\iota-1)} = (X_1, ..., X_{(\iota-1)})$ and determined the variational factors $Q(\vec{\phi}_{(\iota-1)})$, $Q(\vec{Z}_{(\iota-1)})$, $Q(\vec{\mathcal{W}}_{(\iota-1)})$, $Q^{(\iota-1)}(\vec{\alpha})$, $Q^{(\iota-1)}(\vec{\beta})$, $Q^{(\iota-1)}(\vec{\lambda})$, and $Q^{(\iota-1)}(\vec{\tau})$ as well as the parameters $\vec{\pi}^{(\iota-1)}$, $\vec{\eta}^{(\iota-1)}$, $\vec{\epsilon}^{(\iota-1)}$. When the newly arriving data X_ι is observed, we need to update the current ι^{th} optimal value for a variational factor according to the $((\iota-1)^{th})$ values of the other variational factors. Later, we update the ιth optimal value for the second factor by holding the newly obtained ι^{th} value of the first factor fixed and setting other factors still to their $(\iota-1)^{th}$ values. We keep repeating this procedure until all the variational factors are updated with respect to the new observation.

In this work, we first maximize the current lower bound $\mathcal{L}^{(t)}Q$ with respect to $Q(\vec{\phi}_\iota)$, while other variational factors are set to $Q(\vec{Z}_{(\iota-1)})$, $Q(\vec{\mathcal{W}}_{(\iota-1)})$,

$Q^{(\iota-1)}(\vec{\alpha}), Q^{(\iota-1)}(\vec{\beta}), Q^{(\iota-1)}(\vec{\lambda})$, and $Q^{(\iota-1)}(\vec{\tau})$, and the feature saliency $\vec{\epsilon}$ is set to $\vec{\epsilon}^{(\iota-1)}$. Hence, the variational solution for $Q(\vec{\phi}_{\iota})$ can be calculated as

$$Q(\vec{\phi}) = \prod_{l=1}^{D} f_{\iota l}^{\phi_{\iota l}} (1 - f_{\iota l})^{1-\phi_{\iota l}} \tag{8.34}$$

where

$$f_{\iota l} = \frac{\tilde{f}_{\iota l}}{\tilde{f}_{\iota l} + \hat{f}_{\iota l}} \tag{8.35}$$

In Eq. (8.35), we substitute the below values of (8.36) and (8.37) by modifying Eqs. (8.26) and (8.27), respectively, to

$$\tilde{f}_{\iota l} = exp\left\{ \ln \epsilon_{l_1}^{(\iota-1)} + \sum_{j=1}^{M} r_{(\iota-1)j} [\tilde{R}_{jl} + (\overline{\alpha}_{jl} - 1) \ln X_{\iota l} - (\overline{\alpha}_{jl} + \overline{\beta}_{jl}) \ln(1 + X_{\iota l})] \right\} \tag{8.36}$$

$$\hat{f}_{\iota l} = exp\left\{ \ln \epsilon_{l_2}^{(\iota-1)} + \left\{ \sum_{K=1}^{K} m_{(\iota-1)kl} [\tilde{F}_{kl} + (\overline{\lambda}_{kl} - 1) \ln X_{\iota l} - (\overline{\lambda}_{k1} + \overline{\tau}_{k1}) \ln(1 + X_{\iota l})] \right\} \right\} \tag{8.37}$$

In the next step, we maximize the current lower bound $\mathcal{L}^{(t)}Q$ with respect to $Q(\vec{\mathcal{Z}}_{\iota})$, while $Q(\vec{\phi}_{\iota})$ is fixed, $\vec{\pi}$ is set to $\vec{\pi}^{(\iota-1)}$, $Q(\vec{\alpha}), Q(\vec{\beta})$ are set to $Q^{(\iota-1)}(\vec{\alpha}), Q^{(\iota-1)}(\vec{\beta})$, respectively. Based on Eq. (8.18), the variational solution for $Q(\vec{\mathcal{Z}}_{\iota})$ is given by

$$Q(\vec{\mathcal{Z}}_{\iota}) = \prod_{j=1}^{M} \tilde{r}_{\iota j}^{Z_{\iota j}} \tag{8.38}$$

where

$$r_{\iota j} = \frac{\tilde{r}_{\iota j}}{\sum_{j=1}^{M} \tilde{r}_{\iota j}} \tag{8.39}$$

We modify Eq. (8.24) discussed in the previous section to the one below for online variational case

$$\tilde{r}_{\iota j} = exp\left\{ \ln \pi_j{}^{(\iota-1)} + \sum_{l=1}^{D} \left\{ f_{\iota l}\left[\tilde{R}_{jl} + (\overline{\alpha}_{jl} - 1)\ln \mathcal{X}_{\iota l} - (\overline{\alpha}_{jl} + \overline{\beta}_{jl})\ln(1 + \mathcal{X}_{\iota l})\right] \right.\right.$$
$$\left.\left. +(1 - f_{il})\sum_{k=1}^{K} m_{(\iota-1)kl}\left[\tilde{F}_{kl} + (\overline{\lambda}_{kl} - 1)\ln \mathcal{X}_{\iota l} - (\overline{\lambda}_{kl} + \overline{\tau}_{kl})\ln(1 + \mathcal{X}_{\iota l})\right]\right\}\right\} \tag{8.40}$$

Subsequently, we maximize $\mathcal{L}^{(t)}Q$ with respect to $Q(\overrightarrow{\mathcal{W}}_\iota)$, using $Q^{(\iota-1)}(\overrightarrow{\lambda})$, $Q^{(\iota-1)}(\overrightarrow{\tau})$, and $\overrightarrow{\eta}^{(\iota-1)}$, while $Q(\overrightarrow{\phi}_\iota)$ is considered fixed, such that

$$Q(\overrightarrow{\mathcal{W}}_\iota) = \prod_{K=1}^{K}\prod_{L=1}^{D} \mathbf{m}_{\iota\mathbf{kl}}{}^{W_{\iota kl}} \tag{8.41}$$

where

$$m_{\iota kl} = \frac{\tilde{m}_{\iota kl}}{\sum_{k=1}^{k} \tilde{m}_{\iota kl}} \tag{8.42}$$

Equation (8.25) in the previous section is modified as below

$$\tilde{m}_{\iota kl} = exp\left\{ \ln \eta_{kl}^{(\iota-1)} + (1 - f_{(\iota l)})\left[\tilde{F}_{kl} + (\overline{\lambda}_{kl} - 1)\ln X_{\iota l} - (\overline{\lambda}_{kl} + \overline{\tau}_{kl})\ln(1 + X_{\iota l})\right]\right\} \tag{8.43}$$

Now in order to obtain the variational solution for $Q^{(\iota)}(\overrightarrow{\alpha})$, we need to maximize $\mathcal{L}^{(t)}Q$ with respect to the variational factor $Q^{(\iota)}(\overrightarrow{\alpha})$ while holding $Q(\overrightarrow{\phi}_\iota)$ and $Q(\overrightarrow{\mathcal{Z}}_\iota)$ fixed as

$$Q^{\iota}(\overrightarrow{\alpha}) = \prod_{j=1}^{M}\prod_{l=1}^{D} \mathcal{G}\left(\alpha_{jl}^{(\iota)} | u_{jl}^{*(\iota)}, v_{jl}^{*(\iota)}\right) \tag{8.44}$$

A significant characteristic of the adopted variational method [47] which cites that variational inference could be handled as a normal gradient method [49] is that it has convergence to an optimal asymptotic. More precisely, the variational parameter natural gradient is obtained by multiplying the gradient of the parameter with the coefficient matrix inverse (the inverse of the Riemannian metric). This multiplication cancels the coefficient matrix for the distribution of variational posterior parameter and leaves a natural gradient that permits for fast online inference. In this case, the natural gradients of the variational hyper-parameters u^*_{jl} and v^*_{jl} are structurally equivalent to the updates given by

$$\Delta u^{*(\iota)}_{jl} = u^{*(\iota)}_{jl} - u^{*(\iota-1)}_{jl} = u_{j\iota} + N r_{\iota j} f_{\iota j} \overline{\alpha}_{jl} \left[\psi'(\overline{\alpha}_{\iota j} + \tilde{\beta}_{\iota j}) - \psi(\overline{\alpha}_{jl})\right] + \overline{\beta}_{jl}\left[\psi'(\overline{\alpha}_{\iota j} + \tilde{\beta}_{\iota j})\right]\left[\langle \ln \beta_{jl}\rangle - \ln \overline{\beta}_{jl}\right] - u^{*(\iota-1)}_{jl} \tag{8.45}$$

$$\Delta \nu^{*(\iota)}_{jl} = \nu^{*(\iota)}_{jl} - \nu^{*(\iota-1)}_{jl} = \nu_{j\iota} - N r_{\iota j} f_{\iota l} \ln \frac{X_{\iota t}}{1 + X_{\iota t}} - \nu^{*(\iota-1)}_{jl} \tag{8.46}$$

Thus, the variational solutions to hyper-parameters $u^{*(\iota)}_{jl}$ and $v^{*(\iota)}_{jl}$ are calculated through their natural gradients as

$$u^{*(\iota)}_{jl} = u^{*(\iota-1)}_{jl} + \rho_\iota \Delta u^{*(\iota)}_{jl} \tag{8.47}$$

$$\nu^{*(\iota)}_{jl} = \nu^{*(\iota-1)}_{jl} + \rho_\iota \Delta \nu^{*(\iota)}_{jl} \tag{8.48}$$

where ρ_ι is the learning rate and is defined as

$$\rho_\iota = (\delta_o + \iota)^{-\epsilon} \tag{8.49}$$

with the constraints:$\xi \epsilon (0.5, 1]$and $\delta_o \geq 0$. The function of the learning rate here is adopted from [50] and is used to forget the earlier inaccurate estimation of the lower bound and expedite the convergence of the learning process. Online learning embraces the fact that learning environments can (and do) change from second to second. Similarly, the variational factors $Q^{(\iota)}(\vec{\beta})$, $Q^{(\iota)}(\vec{\lambda})$, $Q^{(\iota)}(\vec{\tau})$ are updated as

$$Q^{(\iota)}(\vec{\beta}) = \prod_{j=1}^{M}\prod_{l=1}^{D} \mathcal{G}(\beta^{(\iota)}_{jl} | p^{*(\iota)}_{jl}, q^{*(\iota)}_{jl}) \tag{8.50}$$

$$Q^{(\iota)}(\vec{\lambda}) = \prod_{k=1}^{K}\prod_{l=1}^{D} \mathcal{G}(\lambda^{(\iota)}_{kl} | g^{*(\iota)}_{kl}, h^{*(\iota)}_{kl}) \tag{8.51}$$

$$Q^{(\iota)}(\vec{\tau}) = \prod_{k=1}^{K}\prod_{l=1}^{D} \mathcal{G}(\tau^{(\iota)}_{kl} | s^{*(\iota)}_{kl}, t^{*(\iota)}_{kl}) \tag{8.52}$$

where

$$p^{*(\iota)}_{jl} = p^{*(\iota-1)}_{jl} + \rho_\iota \Delta p^{*(\iota)}_{jl}, \quad q^{*(\iota)}_{jl} = q^{*(\iota-1)}_{jl} + \rho_\iota \Delta q^{*(\iota)}_{jl} \tag{8.53}$$

$$g^{*(\iota)}_{kl} = g^{*(\iota-1)}_{kl} + \rho_\iota \Delta g^{*(\iota)}_{kl}, \quad h^{*(\iota)}_{kl} = h^{*(\iota-1)}_{kl} + \rho_\iota \Delta h^{*(\iota)}_{kl} \tag{8.54}$$

$$s^{*(\iota)}_{kl} = s^{*(\iota-1)}_{kl} + \rho_\iota \Delta s^{*(\iota)}_{kl}, \quad t^{*(\iota)}_{kl} = t^{*(\iota-1)}_{kl} + \rho_\iota \Delta t^{*(\iota)}_{kl} \tag{8.55}$$

The corresponding natural gradients of the variational hyper-parameters in the above equations are given by

$$\Delta p_{jl}^{*(\iota)} = p_{jl}^{*(\iota)} - p_{jl}^{*(\iota-1)} = p_{j\iota} + N r_{\iota j} f_{\iota l} \overline{\beta}_{jl} \big[\psi(\overline{\alpha}_{\iota j} + \overline{\beta}_{\iota j}) - \psi(\overline{\beta}_{jl})\big] + \overline{\alpha}_{jl} \big[\psi'(\overline{\alpha}_{\iota j} + \overline{\beta}_{\iota j})\big] \Big[\langle \ln \alpha_{jl} \rangle - \ln \overline{\alpha}_{jl}\Big] - p_{jl}^{*(\iota-1)} \tag{8.56}$$

$$\Delta q_{jl}^{*(\iota)} = q_{jl}^{*(\iota)} - q_{jl}^{*(\iota-1)} = q_{j\iota} + N r_{\iota j} f_{\iota l} \ln \frac{1}{1 + X_{\iota l}} - q_{jl}^{*(\iota-1)} \tag{8.57}$$

$$\Delta g_{kl}^{*(\iota)} = g_{kl}^{*(\iota)} - g_{kl}^{*(\iota-1)} = g_{kl} + N(1 - \phi_{\iota l}) m_{\iota kl} \overline{\lambda}_{kl} \big[\psi(\overline{\lambda}_{kl} + \overline{\tau}_{kl}) - \psi(\overline{\lambda}_{kl}) + \overline{\tau}_{kl} \big[\psi'(\overline{\lambda}_{kl} + \overline{\tau}_{kl})\Big[\langle \ln \tau_{kl} \rangle - \ln \overline{\tau}_{kl}\Big] - g_{jl}^{*(\iota-1)} \tag{8.58}$$

$$\Delta h_{kl}^{*(\iota)} = h_{kl}^{*(\iota)} - h_{kl}^{*(\iota-1)} = h_{kl} - N(1 - \phi_{\iota l}) m_{\iota kl} \ln \frac{X_{\iota l}}{1 + X_{\iota t}} - h_{kl}^{*(\iota-1)} \tag{8.59}$$

$$\Delta s_{kl}^{*(\iota)} = s_{kl}^{*(\iota)} - s_{kl}^{*(\iota-1)} = s_{kl} + N(1 - \phi_{\iota l}) m_{\iota kl} \overline{\tau}_{kl} \big[\psi(\overline{\lambda}_{kl} + \overline{\tau}_{kl}) - \psi(\overline{\tau}_{kl}) + \overline{\lambda}_{kl \psi'}(\overline{\lambda}_{kl} + \overline{\tau}_{kl})(\langle \ln \lambda_{kl} \rangle - \ln \overline{\lambda}_{kl})\big] - s_{jl}^{*(\iota-1)} \tag{8.60}$$

$$\Delta t_{kl}^{*(\iota)} = t_{kl}^{*(\iota)} - t_{kl}^{*(\iota-1)} = t_{kl} - N(1 - \phi_{kl})) m_{kl} \ln \frac{1}{1 + X_{\iota t}} - t_{kl}^{*(\iota-1)} \tag{8.61}$$

Finally, we can update variational parameters $\vec{\pi}^{(\iota)}$, $\vec{\eta}^{(\iota)}$, and $\vec{\epsilon}^{(\iota)}$ as

$$\pi_j^{(\iota)} = \pi_j^{(\iota-1)} + \rho_\iota \Delta \pi_j \tag{8.62}$$

$$\eta_{kl}^{(\iota)} = \eta_{kl}^{(\iota-1)} + \rho_\iota \Delta \eta_{kl} \tag{8.63}$$

$$\epsilon_{l_1}^{(\iota)} = \epsilon_{l_1}^{(\iota-1)} + \rho_\iota \Delta \epsilon_{l_1}^{\iota} \tag{8.64}$$

where the natural gradients $\Delta \pi_j^{(\iota)}$, $\Delta \eta_{kl}^{(\iota)}$, and $\Delta \epsilon_{l_1}^{(\iota)}$ are calculated by

$$\Delta \pi_j^{(\iota)} = \pi_j^{(\iota)} - \pi_j^{(\iota-1)} = \left(\frac{N}{t}\right) r_{\iota j} - \pi_j^{(\iota-1)} \tag{8.65}$$

$$\Delta \eta_{kl}^{(\iota)} = \eta_{kl}^{(\iota)} - \eta_{kl}^{(\iota-1)} = \left(\frac{N}{t}\right) m_{\iota kl} - \eta_{kl}^{(\iota-1)} \tag{8.66}$$

$$\Delta \epsilon_{l_1}^{(\iota)} = \epsilon_{l_1}^{(\iota)} - \epsilon_{l_1}^{(\iota-1)} = \left(\frac{N}{t}\right) f_{\iota l} - \epsilon_{l_1}^{(\iota-1)} \tag{8.67}$$

The lower bound in the online variational does not always increase even though a new contribution is added to the lower bound for each new observation, whereas it does increase in the case of the batch variational algorithm. Furthermore, as showed in [48], the online variational algorithm can be defined as a stochastic approximation method [51] in order to estimate the expected lower bound and the convergence is assured if the learning standard satisfies these conditions:

$$\sum_{\iota=1}^{\infty} \rho_\iota = \infty, \sum_{i=1}^{\infty} \rho_\iota^2 < \infty \tag{8.68}$$

The major cause of slow convergence is the effect on later estimations due to inaccurate hyper-parameter estimations which occur in the earlier inference stages. Therefore, including the learning rate in the learning process is considered important for accelerating the convergence rate. The steps for online variational inference for the finite GID mixture model with feature selection are abstracted in Algorithm 1. The correct number of components is determined by removing off those with small mixing coefficients close to 0 after convergence. It is worth noticing that although the components number may differ each time a new data is observed, the mixing coefficients are updated in each iteration and hence are suitable to the model framework change automatically.

Algorithm 1: Online variational learning of the finite GID mixture model with feature selection

---

1. Choose the initial number of components M and K.
2. Initialize the values for hyper-parameters $u_{jl}, v_{jl}, p_{jl}, q_{jl}, g_{kl}, h_{kl}, s_{kl}, t_{kl}$.
3. Using K-means algorithm, initialize the values of r_{ij} and m_{ikl}.
4. **for** $t = 1 \rightarrow N$ **do**

   i The variational E-step:
   ii Update the variational solutions for $Q(\vec{\phi}_\iota)$, $Q(\vec{\mathcal{Z}}_\iota)$ and $Q(\vec{\mathcal{W}}_\iota)$ through Eqs. (8.34), (8.38) and (8.41), respectively.
   iii The variational M-step:
   iv Compute learning rate $\rho_\iota = (\delta_o + \iota)^{-\epsilon}$ as in Eq. (8.49)
   v Calculate the natural gradients $\Delta u_{jl}^{*(\iota)}$, $\Delta v_{jl}^{*(\iota)}$, $\Delta p_{jl}^{*(\iota)}$, $\Delta q_{jl}^{*(\iota)}$, $\Delta g_{kl}^{*(\iota)}$, $\Delta h_{kl}^{*(\iota)}$, $\Delta s_{kl}^{*(\iota)}$ and $\Delta t_{kl}^{*(\iota)}$ using Eqs. (8.45),(8.46) , (8.56), (8.57),(8.58), (8.59), (8.60),(8.61), respectively
   vi Update the variational solution for $Q^\iota(\vec{\alpha})$, $Q^{(\iota)}(\vec{\beta})$, $Q^{(\iota)}(\vec{\lambda})$ $Q^{(\iota)}(\vec{\tau})$ through Eqs. (8.44), (8.50), (8.51), (8.52) and (8.61)
   vii Calculate the natural gradients $\Delta\pi_j^{(\iota)}$, $\Delta\eta_{kl}^{(\iota)}$ and $\Delta\epsilon_{l_1}^{(\iota)}$ via Eqs. (8.65), (8.66), (8.67), respectively, for parameters $\vec{\pi}^{(\iota)}$, $\vec{\eta}^{(\iota)}$ and $\vec{\epsilon}^{(\iota)}$
   viii Update the current solutions for $\vec{\pi}^{(\iota)}$, $\vec{\eta}^{(\iota)}$ and $\vec{\epsilon}^{(\iota)}$ using Eqs. (8.62), (8.63), (8.64)
   ix Repeat the variational E-step and M-step until new data is observed.

5. **end for**

8.5 Experimental Results

In this section, we investigate the efficiency of our proposed online variational GID mixture model with feature selection by synthetic data and three challenging medical applications. The synthetic data purpose is to examine the online variational algorithm accuracy in terms of estimation of parameters and model selection. We performed medical image segmentation and feature selection on three data sets of different diseases and different medical image testing techniques. We applied the algorithm to detect brain tumor, skin lesion, and computer-aided detection (CAD) of malaria. Furthermore, we have used two different formats of images to test the applicability of the algorithm on varied output formats, namely MRI scans and dermoscopic photographs. The main goal to focus on medical applications was to visualize the way different analytical and statistical mixture model methods can help the healthcare industry to give more precise results while diagnosing any patient's health using machine learning.

Concerning the medical data sets we have used for the experiments, we make a performance comparison of our algorithm of online variational learning of finite GID with feature selection (OVGIDMM) with two other models, namely online variational learning of finite inverted Dirichlet mixture model (OVIDMM) and online variational learning of finite Gaussian mixture model (OVGMM) to illustrate the merits of our algorithm implementation. OVIDMM is considered for comparison since it has less covariance compared to our proposed algorithm and can also be used for positive vectors. OVGMM is considered since it is an extremely popular and novel approach. The below sections would follow the description on image segmentation, feature selection, and the results obtained by calculating different evaluation metrics.

8.5.1 Image Segmentation

Image segmentation is considered an integral part for computer vision. It is the process used to partition the image into many segments according to the pixels. The main aim of the segmentation process is to change the representation of the image to make the analysis and interpretation process easier, since we get more understanding about the image and to detect the lines or curves in the image. In other words, the image segmentation makes a label for each pixel in order to have a table of similar features. Each pixel is similar to the other computed features like color or texture.

There are mainly two types of image segmentation techniques called non-contextual thresholding and contextual thresholding. The non-contextual type does not consider the spatial relationships between features in the image but the contextual technique does consider these relationships, for example, grouping together pixels with similar gray levels. In all our experiments in this chapter, we used the non-contextual thresholding technique called RGB color thresholding. The input to the thresholding operation was typically grayscale for brain tumor detection and color scale for the skin melanoma and CAD of malaria. In this implementation, the output is a binary image representing the segmentation where the black pixels correspond to the background and white pixels correspond to the foreground (or vice versa). The detection of edges in various clusters formed by the image segmentation helped us to derive the diagnostic insights to it by comparing it with the ground truth. The major challenge while performing segmentation was to identify the pixels that belong to features of interest to us. As an example, we performed the following steps to detect the brain tumor by MRI image segmentation: where we first extracted the brain structures and then did localization of tumor region of interest (ROI) and then considering the size of the tumor with other structures in the brain and then diagnosed the tumor by comparing it to the ground truth.

In automated MRI image analysis, image segmentation is considered to be a preliminary step. The different types of factors which can affect on deciding the segmentation type when dealing with medical data sets are: which main body part is being considered, the imaging technique, and lastly the application type for deciding the best suitable segmentation [52]. The applications in the healthcare field could be related to cell counting, measurement process for organ, counting of cells, or prediction of abnormal growth which would depend on boundary extraction. There are a few general challenges that could be experienced when dealing with medical image segmentation: (1) the variability in sensing of the main part is very large, especially because it is very complicated when dealing with human anatomy, (2) the effect of medical image is different for each organ of the body since the motion of the heart also affects the imaging quality, (3) the noise effect of the sensor being used for detection.

In our model we extracted the feature of each image using the most commonly used technique of color histogram where we calculated the green color component histogram value for an RGB component of an image since the red and blue color component had no variations and followed no statistical model. Color is one of the most outstanding features of the image, it is the most important human visual content, and it is very easy to calculate. The color histogram for an image is constructed by quantizing the colors within the image and counting the number of pixels of each color. Then, we take a summation of it and find the mean and standard deviation from the color histogram. Finally, it is stored in a 1D array. This value is calculated for every image in the data set [53]. The graph in Fig. 8.2 shows two green component value color histogram graph which has been obtained by running our algorithm on malaria data set classifying the images into two categories uninfected and parasitized, respectively, for all the patients in the data set.

8.5.2 Synthetic Data

Our proposed algorithm was evaluated by quantitative analysis on dimensional data with two relevant features. These data sets have different data sizes, namely 200,

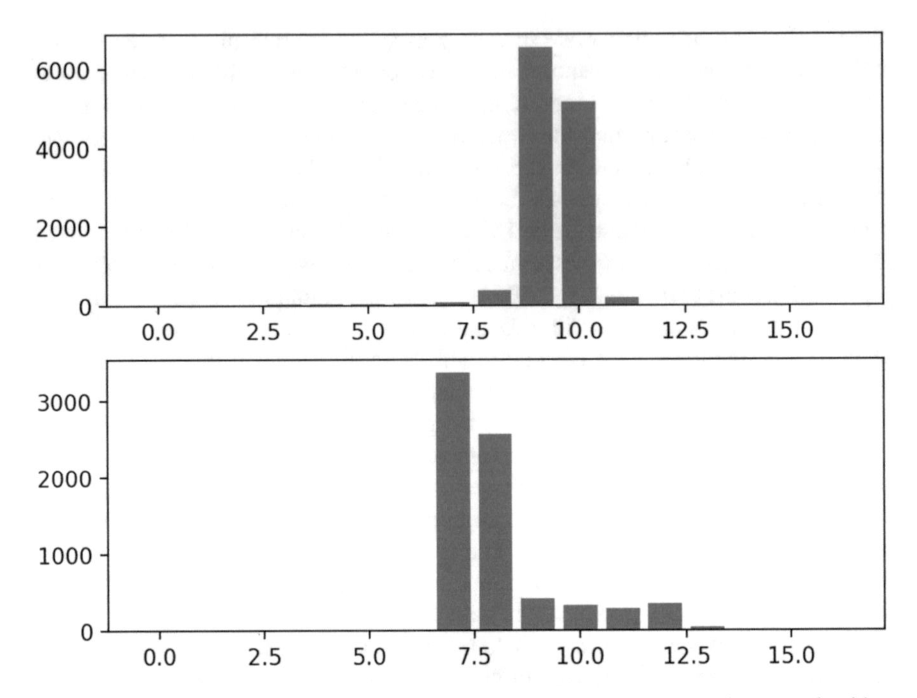

Fig. 8.2 Color histogram for green component in RGB obtained by running our algorithm OVGIDMM on CAD of malaria data set. The top graph shows the uninfected patients and the bottom parasitized patients

600, 900, and 1200. The relevant features were created in the converted space from a mixture of inverted beta distributions with well-separated components. Table 8.1 demonstrates the actual and evaluated parameters of the distributions using our proposed online variational approach and considering the relevant features for each data set. According to the results obtained, the model parameters represent relevant features, and its mixing coefficients are precisely estimated by our online algorithm. In our experiments for synthetic data, the components M and K number had been initialized with 6 and 2 for two-dimensional data, respectively, with equivalent mixing coefficients and the feature salencies value is initialized at 0.5. The initial values of the hyper-parameters u, p, g, and s for the conjugate priors are fixed to 1, v to 0.04, q to 0.03, h to 0.05, and t to 0.06. ϵ and Σ the learning rate parameters are fixed to 0.5 and 64.

There are two diverse methods that can be used for estimating the correct number of components. Firstly, it can be done by directly applying our proposed online algorithm to calculate the mixing coefficients. Thus, the correct number of components was considered by eliminating the components with very small (less than 10^{-5}) mixing coefficients in each data set. The second method is to exploit the variational likelihood bound as a model selection score to find the appropriate

Table 8.1 Real and estimated parameters of different data sets

Data set	N_j	j	α_{j1}	β_{j1}	α_{j2}	β_{j2}	π_j	$\hat{\alpha_{j1}}$	$\hat{\beta_{j1}}$	$\hat{\alpha_{j2}}$	$\hat{\beta_{j2}}$	$\hat{\pi}_j$
Data set 1	100	1	20	13	18	15	0.5	19.91	13.05	18.99	15.30	0.50
($N = 200$)	100	2	25	15	22	12	0.5	25.43	15.27	21.76	12.27	0.50
Data set 2	200	1	20	13	24	15	0.33	21.59	14.53	23.74	15.13	0.33
($N = 600$)	200	2	22	15	25	12	0.33	21.19	15.11	25.41	12.76	0.33
	200	3	25	16	22	14	0.34	24.79	16.10	23.86	13.86	0.34
Data set 3	300	1	20	13	24	15	0.33	20.68	13.51	24.17	14.13	0.33
($N = 900$)	300	2	22	15	25	12	0.33	21.89	14.77	24.46	13.26	0.33
	300	3	21	15	22	14	0.34	20.53	15.05	22.82	13.89	0.34
Data set 4	400	1	20	13	20	15	0.33	20.06	14.37	21.49	13.88	0.33
($N = 1200$)	400	2	22	15	20	12	0.33	21.72	14.96	20.83	14.21	0.33
	400	3	21	15	22	14	0.34	20.89	13.89	23.02	14.28	0.34

N denotes the total number of data points, N_j denotes the number of data points in the cluster $j.\alpha_{j1}, \beta_{j1}, \alpha_{j2}, \beta_{j2}$, and π_j are the real parameters and $\hat{\alpha}_{j1}, \hat{\beta}_{j1}, \hat{\alpha}_{j2}, \hat{\beta}_{j2}$, and $\hat{\pi}_j$ are the parameters estimated by our proposed algorithm

number of components. Especially, we fixed the number of components M and calculated the lower bound.

8.5.3 *Medical Image Data Sets*

After validating the algorithm on synthetic data sets, we applied it on three challenging medical data sets for brain tumor detection, skin melanoma detection, and CAD of malaria data set. We observed that our algorithm could detect the morphological and structural anomalies similar to the ground truth data when performing image segmentation. We used 30 different patient images in each case of image segmentation for brain tumor detection and skin melanoma and compared the result of our proposed algorithm OVGIDMM with OVIDMM and OVGMM by taking out the mean of all the images values obtained in terms of adjusted Rand index (ARI) score, adjusted mutual information (AMI) score, V-measure score, Fowlkes–Mallows (FM) index, Dice similarity coefficient, and Jaccard similarity index for evaluation of the accuracy. The evaluation of CAD of malaria data set was also done by comparing our algorithm to OVIDMM and OVGMM on the basis of confusion matrix for classification of the patients into uninfected and parasitized category.
In our experiments for image segmentation of brain MRI images Sect. 8.5.3.1 and skin melanoma images Sect. 8.5.3.2, the number of components M and K had been initialized with 16 and 2 and the feature salencies value was initialized at 0.5. The initial values of the hyper-parameters u, p, g, and s for the conjugate priors are fixed to 1, v to 0.03, q to 0.035, s to 0.05, and t to 0.06. ϵ and Σ the learning

rate parameters are fixed to 0.5 and 64. The initialization was kept different for the testing of CAD of malaria data set which has been described in Sect. 8.5.3.3.

8.5.3.1 Brain Tumor Detection

Basically, a tumor is an uncontrolled growth of cancerous cells that can happen in any part of the body, while the brain tumor is an uncontrolled growth of cancerous cells happening in the brain itself. Tumors of brain could be benign or malignant. The benign brain tumor structure has uniformity and does not include active (cancer) cells, while in the case of malignant brain tumors, the tumor structure has non-uniformity (heterogeneous) and include active cells [54]. Examples of low-grade tumors are gliomas and meningiomas which are labeled as benign tumors. A class of high-grade tumors of astrocytomas is called as glioblastoma which is labeled as malignant tumors.

As per the World Health Organization and American Brain Tumor Association [55], the most common grading system to classify brain tumors into benign and malignant is by knowing the cell origin and how the cells behave, from the least aggressive (benign) to the most aggressive (malignant). Some tumor types are assigned a grade, ranging from grade I (least malignant) to grade IV (most malignants), which signifies the rate of growth of the tumor. According to that standards, benign tumors come under grade I and II glioma and malignant tumors come under grade III and IV glioma. The grade I and II glioma are also labeled low-grade tumor kind and have slow growth, while grade III and IV are labeled high-grade tumor kinds and have a rapid growth. If the brain tumor of low-grade is left uncured, there is a high risk for it to get developed into a tumor of high-grade malignant tumor.

The low-grade I and II glioma benign tumors are deemed to be therapeutic below full surgical excursion, while grade III and IV malignant brain tumors category can be cured via radiotherapy, chemotherapy, or a combination of these. The malignant glioma category includes together grade III and IV gliomas, which is also indicated to as anaplastic astrocytomas. A mid-grade tumor is an anaplastic astrocytoma which shows abnormal, irregular growing, and an expanded growth indication compared to other low-grade tumors. Glioblastoma is the common malignant compose of astrocytoma. The blood vessels unnatural fast growing and of the necrosis existence (dead cells) round the tumor are characterized glioblastoma from the entire other grades of the tumor class. The class of grade IV tumor which is glioblastoma is always fast growing and has an extremely malignant structure of tumors as compared to the other tumor grades. Several researchers have proposed various methodologies and algorithms for brain tumor segmentation by using K-means clustering technique [56], Spatial Fuzzy C-means [57], convolution neural network (CNN) as pixel classifier for the segmentation process [58], and K-Medoids clustering [59].

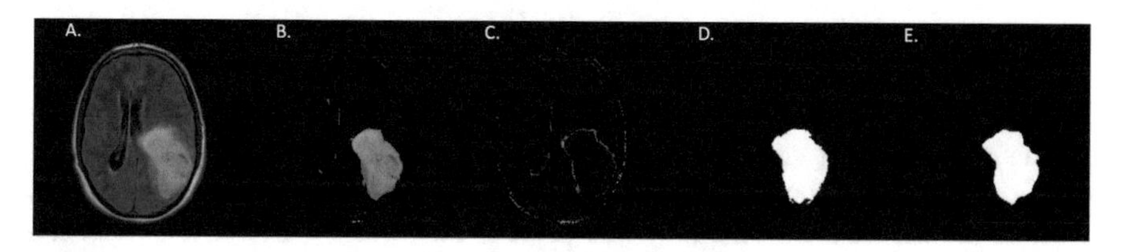

Fig. 8.3 Example of best segmented brain MRI images for patient 1: (**a**) input MRI image, (**b**) 7th cluster image, (**c**) 8th cluster image, (**d**) predicted image from post-processing, (**e**) ground truth image

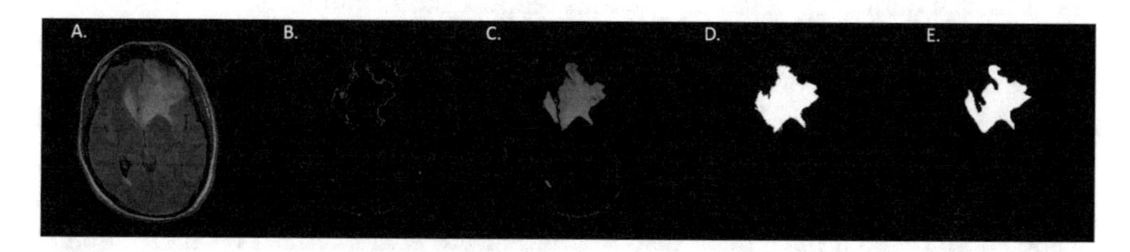

Fig. 8.4 Example of best segmented brain MRI images for patient 2: (**a**) input MRI image, (**b**) 0th cluster image, (**c**) 5th cluster image, (**d**) predicted image from post-processing, (**e**) ground truth image

Table 8.2 Evaluation metrics for brain tumor detection

Method	ARI	AMI	V-Measure	Dice	Jaccard
OVGIDMM	**90.44**	**78.66**	**80.80**	**91.12**	**82.97**
OVIDMM	84.02	67.9	72.11	86.38	75.0
OVGMM	82.3	65.83	70.90	83.63	73.02

In this chapter, the brain tumor data set was obtained from kaggle.[1] The data set consisted of 110 brain MRI images in the FLAIR sequence along with manual FLAIR abnormality segmentation masks which are binary, 1-channel images considered as ground truth. The images for the data set have been obtained from The Cancer Imaging Archive (TCIA). In order to find out the brain tumor from modalities of the brain MRI images, image segmentation was performed along with some post-processing steps. The representative segmentation achieved after running our proposed algorithm is depicted in Figs. 8.3 and 8.4 for two different patients as an example where two of the best segmented clusters generated by the algorithm are merged as a post-processing step in order to compare with the ground truth.

Table 8.2 shows the performance comparison between our proposed algorithm, OVIDMM and OVGMM. The result obtained from our algorithm is clearly much better in terms of accuracy for all evaluation metrics compared to OVIDMM and OVGMM signifying our algorithm could be of better use in healthcare to diagnose brain tumor.

[1] https://www.kaggle.com/mateuszbuda/lgg-mri-segmentation.

8.5.3.2 Skin Melanoma Detection

Melanoma is a type of skin cancer which generates from melanocytes (melanin-producing cells) that are naturally existed in the epidermis basal layer or inside the dermis. It may grow in developed way inside a present naevus of melanocytic. Despite the fact that melanoma is less popular than non-melanoma skin cancers, namely squamous cell carcinoma and basal cell carcinoma it causes noticeably more cases of death. Melanoma can be curable if it is diagnosed when it is confined to the outer layers of the skin. Therefore, the early detection of these always makes it easier for the clinicalogist to help patients recover soon from it.

Melanoma can happen anywhere on the skin and on unique occasions can also be present in other tissues such as the eye, central nervous system, gastrointestinal mucosa, genitourinary and respiratory systems. It can also be classified into benign and malignant. It is extremely difficult to differentiate early stage melanoma from benign skin lesions due to the similarity in structure. There are five principal kinds of melanoma which have been discovered and are characterized by their clinical looks, the growing lesion behavior, and their position on our body. There are ten of the melanomas that are approximately non-pigmented (amelanotic), which raises the toughness in making a correct diagnosis for the patients. Adding to the five principle kinds of melanomas there are a number of other more unusual kinds, such as desmoplastic melanoma and malignant blue naevus. This type of cancer is primarily diagnosed visually. After initial clinical screening and dermoscopic analysis, a biopsy and histopathological sample is analyzed.

Even when experienced dermatologists use dermoscopy for diagnosis, computer-aided diagnosis (CAD) is useful for increasing the accuracy and speed of diagnostics. Dermoscopy based on computerized non-invasive dermatology thus is getting very essential for physicians to inspect the pigmented skin lesions and detect malignant melanoma at an early stage to reduce cost and improve the survival rate. Therefore, a CAD system is required to segment melanomas as precisely as possible and also investigate the dermoscopic images. Though, skin lesion segmentation in dermoscopic images is a challenge due to their blurry and irregular boundaries [60]. The computer is no smarter than a human being, but it can extract some information, such as color variation, asymmetry, and plot characteristics, which may not be readily apparent to human eyes. Many systems and algorithms have been proposed, such as the seven-point checklist, the ABCD rule, and the Menzies method to improve the diagnosis of cutaneous melanoma carcinoma.

In order to test the performance of our algorithm to detect skin melanoma we used the data set from International Skin Imaging Collaboration.[2] The data set consists of 23,906 dermoscopic images of melanoma of different patients with ground truth available for each image. Figures 8.5 and 8.6 are example images of two different patients, respectively, showing the result of our proposed algorithm while performing image segmentation with feature selection. In each patient's

[2]https://www.isic-archive.com.

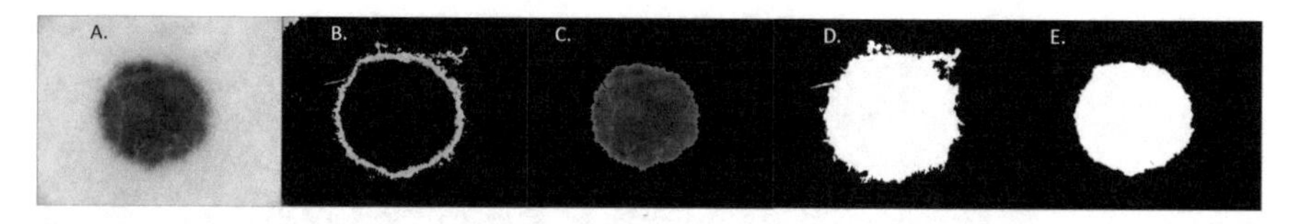

Fig. 8.5 Example of best segmented dermoscopic images for patient 1: (**a**) input image, (**b**) 5th cluster image, (**c**) 8th cluster image, (**d**) predicted image from post-processing, (**e**) ground truth image

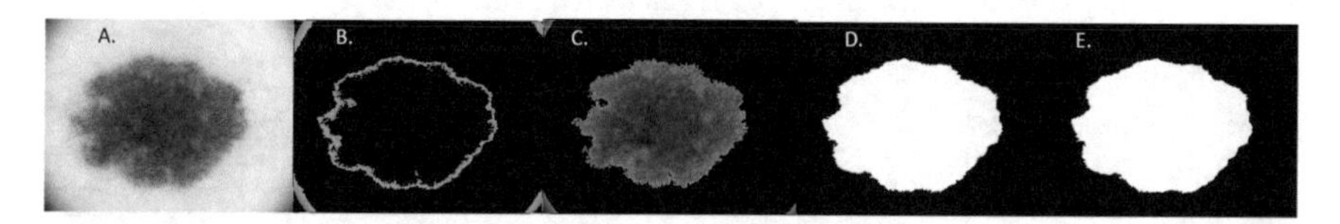

Fig. 8.6 Example of best segmented dermoscopic images for patient 2: (**a**) input image, (**b**) 0th cluster image, (**c**) 5th cluster image, (**d**) predicted image from post-processing, (**e**) ground truth image

Table 8.3 Evaluation metrics for skin melanoma detection

Method	ARI	AMI	V-Measure	FM	Dice	Jaccard
OVGIDMM	**88.22**	**75.04**	**78.56**	**94.59**	**95.95**	**92.48**
OVIDMM	75.46	62.43	67.03	89.70	89.67	82.17
OVGMM	50.57	42.86	46.04	82.32	69.90	58.81

case there were a lot of cluster images formed up to approximately the number of components however, in post-processing of the image we merged the best segmented images in order to compare it with ground truth.

Table 8.3 shows the result obtained from our algorithm, as compared to OVIDMM and OVGMM which is much accurate for all evaluation metrics where we took the mean of the test performed on 30 sample images.

8.5.3.3 Malaria Data Set

Malaria is a disease which can transmit from person to another through infected mosquitoes. In 2015, the World Health Organization (WHO) estimated that 438,000 people have dead because of malaria; the Institute of Health Metrics and Evaluation (IHME), Global Burden of Disease (GBD) set this estimations at 620,000 in 2017. The statistics prove that malaria can be a very fatal disease if not diagnosed on time. It should be diagnosed quickly for treating the patient on time and to prohibit further spreading of infection in the society by local mosquitoes. In order to avoid the issues of not getting timely diagnosis and error prone detection, computer-aided detection (CAD) and mathematical morphology are applied as effective tools for computer-aided malaria detection and classification. These techniques are widely

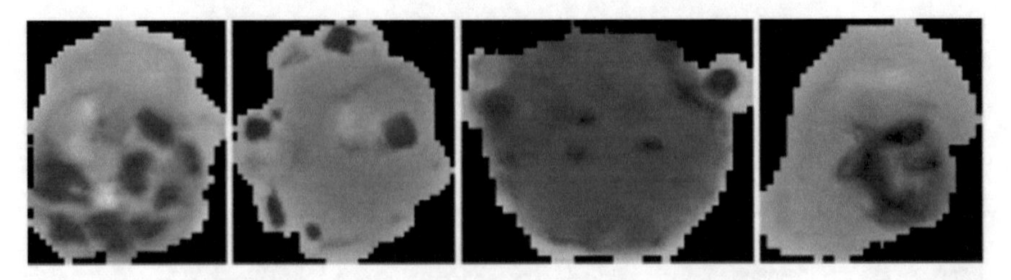

Fig. 8.7 Examples of malaria cells labeled as parasitized in the data set

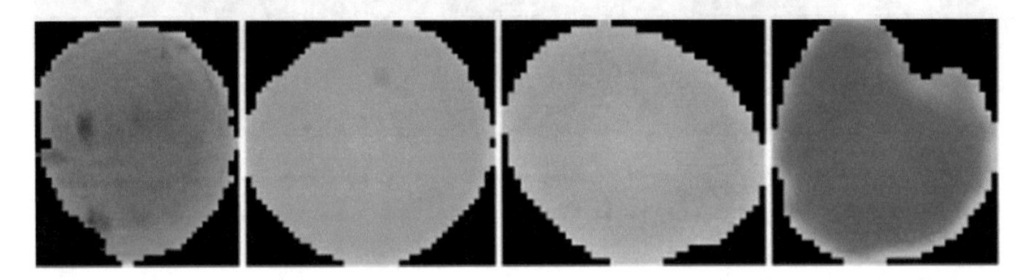

Fig. 8.8 Examples of malaria cells labeled as uninfected in the data set

used for image processing purposes and employed successfully in biomedical image analysis.

The infection of malaria parasite causes microstructural changes to the erythrocytes. The RBCs microscopic features are usually specified to morphology, intensity, and texture. Also, they may perform the differences that happen between healthy and unhealthy cells. Both textural and geometric merits for demonstrating stages of malaria infection have been reported in most of the studies. In general, merits may be identified according to the next characteristics: morphological features and textural and intensity features [61]. It is a popular arithmetical morphology procedure to compute the grains size distribution in binary images, by a sequence of morphological opening operations. Some authors utilize the area granulometry for prepossessing goals in malaria description, although it is certainly efficient for extracting cell size features. Local area granulometry connected with color histogram is employed as features. The feature of area granulometry is computed locally on the stained objects binary mask, for channels of RGB.

In this chapter, we used the malaria data set from NIH.[3] The data set includes a sum of 27,558 cell images with equivalent examples of parasitized and uninfected cells. A few examples of the images from the data set are illustrated in Fig. 8.7 of parasitized cells and Fig. 8.8 of uninfected cells. The data set also includes a csv file including the Patient-ID to cell mappings for the parasitized and uninfected classes. There are 151 patient entries for the parasitized class and the uninfected class includes 201 entries as the normal cells. In our experiments, the feature selection

[3]https://ceb.nlm.nih.gov/repositories/malaria-datasets/.

Table 8.4 Evaluation metrics for malaria data set

Method	Accuracy	Precision	Recall	F1-score	Estimation time (s)
OVGIDMM	**93.8**	**95.10**	**90.06**	**92.51**	**0.3**
OVIDMM	90.3	87.96	79.47	87.59	0.5
OVGMM	83.80	75.26	92.71	83.08	1.2

concept played a very crucial role in this data set to evaluate the performance of our algorithm. The features were extracted using the color histogram method where we considered specifically the green component of RGB model since the red and blue had no variations. The same has been described in Sect. 8.5.1. Feature extraction has the target of decreasing the subsequent computational complication and facilitating a credible and accurate recognition for unknown new data. For this experiment, the number of components M and K had been initialized with 2 and 4 and the feature salencies value was initialized at 0.5. The initial values of the hyper-parameters u, p, g, v, and s for the conjugate priors are fixed to 1, q, h, and t were set to 10. ϵ and Σ the learning rate parameters are fixed to 0.5 and 64. In total, we considered 17 features out of which 4 were considered as relevant and the rest as irrelevant.

Table 8.4 shows precisely the way our proposed algorithm outperforms the other two algorithms in CAD of malaria by giving greater accuracy as well as taking less time for execution. It also proves the fact that it takes less time for convergence as compared to the other two models.

8.6 Conclusion

This chapter introduces an online variational algorithm for finite GID with an unsupervised feature selection method with the main focus on health sector. The use of data in healthcare has been attributed to escalating highly in the recent past [62]. As a matter of fact, the doctor's written notes in the current society have been converted to electronic records. Similarly, it has been of prime importance to have efficient CAD systems for accurate diagnosis of diseases and increasing survival rate. This has facilitated the dedication of researching medical data mining, computer vision, and data processing for effective current medical data practices. The aim of this revolution in the medical sector has been to realize an effective way of reducing treatment costs and improving efficiency in patient handling techniques.

In the proposed improved learning framework we could detect brain tumor, skin melanoma, and malaria with high accuracy compared to the other tested models of OVIDMM and OVGMM. In our proposed algorithm the complexity of mixture model is identified concurrently with the parameters estimation and the features selection. The learning process is based upon variational inference in an online manner and permitting closed-form solutions for the various involved model parameters. The algorithm also proved the crucial role of online variational learning where data becomes available in a sequential order, thus sequentially updating the best predictor for the future data at each step, as opposed to batch learning technique which generates the best predictor by learning the entire data set at once. The online variational learning as an extension to classic variational inference method keeps not only the advantages of previous models, but also speeds up the convergence rate significantly which has been observed in the malaria data set when handling large data.

The future work can be devoted to have an extension of online model to finite inverted beta Liouville mixture model, which would help us improve the model learning and have better results.

Appendix

Update the Variational Hyper-Parameters

Instead of using the gradient method to estimate u^*_{jl} and v^*_{jl}, for the variational factor $Q(\alpha_{jl})$, it is more straightforward to apply Eq. (8.17) for computing the variational solutions. Therefore, the logarithm of $Q(\alpha_{jl})$ is given by

$$\ln Q(\alpha_{jl}) = \big\langle \ln p\big(\mathcal{X}, \Lambda \mid \Upsilon\big)\big\rangle_{\Lambda \neq \alpha_{jl}} = \sum_{i=1}^{N} \langle z_{(ij)}\rangle\langle \phi_{(il)}\rangle D(\alpha_{jl})$$

$$+\alpha_{jl} \sum_{i=1}^{N} \langle z_{(ij)}\rangle\langle \phi_{(il)}\rangle + \ln_{il}(u_{jl} - 1) \ln \alpha_{il} + \nu_{jl}\alpha_{jl} + \text{const} \tag{8.69}$$

where we have defined $D(\alpha_{jl})$ as

$$D(\alpha_{jl}) = \left\langle \frac{\Gamma\big(\alpha_{jl} + \beta_{jl}\big)}{\Gamma\big(\alpha_{jl}\big)\Gamma\big(\beta jl\big)} \right\rangle_{\beta_{jl}} \tag{8.70}$$

Due to the intractability of the function $D(\alpha_{jl})$, variational inference cannot be performed directly and Eq. (8.69) does not have the same form as the logarithm of a Gamma distribution as its conjugate prior. This problem is tackled by exploiting a non-linear approximation of the function $D(\alpha)$ as proposed in [43].

$$D(\alpha) \geq \ln \alpha \Big[\psi\big(\overline{\alpha} + \tilde{\beta}\big) - \psi\big(\overline{\alpha}\big)\Big] + \beta\psi\big(\overline{\alpha} + \tilde{\beta}\big)[\langle \ln \beta\rangle - \ln \overline{\beta}]\overline{\alpha} \tag{8.71}$$

After substituting the lower bound (8.71) back into (8.69), we obtain

$$\ln Q(\alpha_{jl}) \approx \ln \alpha_{jl} \Bigg\{ \sum_{l=1}^{N} \langle Z_{il} \rangle \langle \phi_{il} \rangle \Big[\psi(\overline{\alpha}_{jl}) + (\tilde{\beta}_{jl}) - \psi(\overline{\alpha}_{jl}) \Big]$$

$$+\beta \psi'(\overline{\alpha} + \tilde{\beta}) [\langle \ln \beta \rangle - \ln \overline{\beta}] \overline{\alpha} - (u_{jl} - 1) \Bigg\}$$

$$+\alpha_{jl} \Bigg\{ \sum_{i=1}^{N} \langle z_{(ij)} \rangle \langle \phi_{(il)} \rangle \ln X_{il} - \nu_{jl} \Bigg\} + const \tag{8.72}$$

Noticeably (8.72), has the logarithmic form of a Gamma distribution. By taking the exponential of both sides of (8.72), we can obtain

$$Q(\vec{\alpha}) = \prod_{j=1}^{M} \prod_{l=1}^{D} \mathcal{G}(\alpha_{jl} | u_{jl}^{*}, \nu_{jl}^{*}) \tag{8.73}$$

where the hyper-parameters u_{jl}^{*}andv_{jl}^{*} can be estimated as:

$$u_{jl}^{*} = u_{jl} + \sum_{l=1}^{N} \langle Z_{il} \rangle \langle \phi_{il} \rangle \overline{\alpha}_{jl} \Big[\psi(\overline{\alpha}_{jl} + \overline{\beta}_{jl}) - \psi(\overline{\alpha}_{jl}) \Big] \tag{8.74}$$

$$+\beta \psi'(\overline{\alpha}_{jl} + \overline{\beta}_{jl}) [\langle \ln \beta_{jl} \rangle - \ln \overline{\beta}_{jl}] \tag{8.75}$$

$$\nu_{jl}^{*} = \nu_{jl} + \sum_{l=1}^{N} \langle Z_{il} \rangle \langle \phi_{il} \rangle \ln \frac{X_{il}}{1 + X_{il}} \tag{8.76}$$

Since $\alpha, \beta,$ and τ all have Gamma priors it is straightforward to obtain the variational solutions for $Q(\alpha)$, $Q(\beta)$, and $Q(\tau)$ in a same procedure as for $Q(\lambda)$. These variational solutions for hyper-parameters are

$$p_{jl}^{*} = p_{jl} + \sum_{l=1}^{N} \langle Z_{il} \rangle \langle \phi_{il} \rangle \overline{\beta}_{jl} \Big[\psi(\overline{\alpha}_{jl} + \tilde{\beta}_{jl}) - \psi(\overline{\beta}_{jl}) \Big]$$

$$+\overline{\alpha}_{jl} [\psi'(\overline{\alpha}_{jl} + \overline{\beta}_{jl})] \Big[\langle \ln \alpha_{jl} \rangle - \ln \overline{\alpha}_{jl} \Big] \tag{8.77}$$

$$q_{jl}^{*} = q_{jl} - \sum_{l=1}^{N} \langle Z_{il} \rangle \langle \phi_{il} \rangle \ln \frac{1}{1 + X_{il}} \tag{8.78}$$

$$g_{kl}^* = g_{kl} + \sum_{l=1}^{N} \langle W_{ikl} \rangle \langle (1 - \phi_{il}) \rangle \overline{\lambda}_{kl} \big[\psi(\overline{\lambda}_{kl} + \overline{\tau}_{kl})$$

$$-\psi(\overline{\lambda}_{kl}) + \overline{\tau}_{kl} \big[\psi'(\overline{\lambda}_{kl} + \overline{\tau}_{kl}) \Big[\langle \ln \tau_{kl} \rangle - \ln \overline{\tau}_{kl} \Big] \tag{8.79}$$

$$h_{kl}^* = h_{kl} + \sum_{l=1}^{N} \langle W_{ikl} \rangle \langle (1 - \phi_{il}) \rangle \ln \frac{\mathcal{X}_{il}}{1 + \mathcal{X}_{il}} \tag{8.80}$$

$$s_{kl}^* = s_{kl} + \sum_{l=1}^{N} \langle W_{ikl} \rangle \langle (1 - \phi_{il}) \rangle \overline{\tau}_{kl} \big[\psi(\overline{\lambda}_{kl} + \overline{\tau}_{kl})$$

$$-\psi(\overline{\tau}_{kl}) + \overline{\lambda}_{kl\psi'}(\overline{\lambda}_{kl} + \overline{\tau}_{kl})(\langle \ln \lambda_{kl} \rangle - \ln \overline{\lambda}_{kl}) \big] \tag{8.81}$$

$$t_{kl}^* = t_{kl} + \sum_{l=1}^{N} \langle W_{ikl} \rangle \langle (1 - \phi_{il}) \rangle \ln \frac{1}{1 + \mathcal{X}_{il}} \tag{8.82}$$

where

$$\langle Z_{ij} \rangle = r_{ij}, \quad \langle w_{ikl} \rangle = m_{ikl}$$

$$\langle \phi_{ij} \rangle = f_{ij}, \quad \langle (1 - \phi_{il}) \rangle = (1 - f_{il})$$

References

1. H.P. Ng, Sim Ong, Kelvin Foong, Poh-Sun Goh, and Wieslaw Nowinski. Medical image segmentation using k-means clustering and improved watershed algorithm. *Proceedings of the IEEE Southwest Symposium on Image Analysis and Interpretation*, 2006:61–65, 02 0001.
2. Zhensong Chen, Zhiquan Qi, Fan Meng, Limeng Cui, and Yong Shi. Image segmentation via improving clustering algorithms with density and distance. *Procedia Computer Science*, 55:1015–1022, 2015. 3rd International Conference on Information Technology and Quantitative Management, ITQM 2015.
3. A Ajala Funmilola, OA Oke, TO Adedeji, OM Alade, and EA Adewusi. Fuzzy kc-means clustering algorithm for medical image segmentation. *Journal of Information Engineering and Applications, ISSN*, 22245782:2225–0506, 2012.
4. Atienza N., García-Heras J., Muñoz-Pichardo J.M., and Villa R. An application of mixture distributions in modelization of length of hospital stay. *Statistics in Medicine*, 27(9):1403–1420, 2008.
5. Meeta Kalra, Michael Osadebey, Nizar Bouguila, Marius Pedersen, and Wentao Fan. *Online Variational Learning for Medical Image Data Clustering*, pages 235–269. Springer International Publishing, 2020.

6. Wenmin Chen, Wentao Fan, Nizar Bouguila, and Bineng Zhong. *Medical Image Segmentation Based on Spatially Constrained Inverted Beta-Liouville Mixture Models*, pages 307–324. Springer International Publishing, Cham, 2020.
7. R. Xu and D. C. Wunsch. Clustering algorithms in biomedical research: A review. *IEEE Reviews in Biomedical Engineering*, 3:120–154, 2010.
8. Trevor Hastie and Robert Tibshirani. Discriminant analysis by Gaussian mixtures. *Journal of the Royal Statistical Society, Series B*, 58:155–176, 1996.
9. Mandar Dixit, Nikhil Rasiwasia, and Nuno Vasconcelos. Adapted Gaussian models for image classification. In *CVPR 2011*, pages 937–943. IEEE, 2011.
10. Taoufik Bdiri and Nizar Bouguila. Bayesian learning of inverted Dirichlet mixtures for svm kernels generation. *Neural Computing and Applications*, 23(5):1443–1458, 2013.
11. Ahmed Elgammal, David Harwood, and Larry Davis. Non-parametric model for background subtraction. In *European conference on computer vision*, pages 751–767. Springer, 2000.
12. HS Kuyuk, E Yildirim, E Dogan, and G Horasan. Application of k-means and Gaussian mixture model for classification of seismic activities in Istanbul. *Nonlinear Processes in Geophysics*, 19(4):411–419, 2012.
13. Tarek Elguebaly and Nizar Bouguila. A Bayesian approach for sar images segmentation and changes detection. In *2010 25th Biennial Symposium on Communications*, pages 24–27. IEEE, 2010.
14. S. Boutemedjet, N. Bouguila, and D. Ziou. A hybrid feature extraction selection approach for high-dimensional non-Gaussian data clustering. *IEEE Transactions on Pattern Analysis and Machine Intelligence*, 31(8):1429–1443, Aug 2009.
15. Wentao Fan and Nizar Bouguila. Online variational learning of generalized Dirichlet mixture models with feature selection. *Neurocomputing*, 126:166–179, 2014. Recent trends in Intelligent Data Analysis Online Data Processing.
16. Taoufik Bdiri and Nizar Bouguila. Learning inverted Dirichlet mixtures for positive data clustering. In *International Workshop on Rough Sets, Fuzzy Sets, Data Mining, and Granular-Soft Computing*, pages 265–272. Springer, 2011.
17. Taoufik Bdiri, Nizar Bouguila, and Djemel Ziou. Variational Bayesian inference for infinite generalized inverted Dirichlet mixtures with feature selection and its application to clustering. *Applied Intelligence*, 44(3):507–525, Apr 2016.
18. Sami Bourouis, Mohamed Al Mashrgy, and Nizar Bouguila. Bayesian learning of finite generalized inverted Dirichlet mixtures: Application to object classification and forgery detection. *Expert Systems with Applications*, 41(5):2329–2336, 2014.
19. Wentao Fan, Nizar Bouguila, and Djemel Ziou. Variational learning of finite Dirichlet mixture models using component splitting. *Neurocomputing*, 129:3–16, 2014.
20. Robert Christian and George Casella. Monte Carlo statistical methods (book review). *Technometrics*, 42(4):430, 1999.
21. Michael I Jordan, Zoubin Ghahramani, Tommi S Jaakkola, and Lawrence K Saul. An introduction to variational methods for graphical models, learning in graphical models, 1999.
22. Zoubin Ghahramani and Matthew J Beal. Variational inference for Bayesian mixtures of factor analysers. In *Advances in neural information processing systems*, pages 449–455, 2000.
23. Cedric Archambeau and Manfred Opper. Approximate inference for continuous-time Markov processes. *Bayesian Time Series Models*, pages 125–140, 2011.
24. Manfred Opper and Guido Sanguinetti. Variational inference for Markov jump processes. In *Advances in neural information processing systems*, pages 1105–1112, 2008.
25. Wentao Fan and Nizar Bouguila. Online variational learning of finite Dirichlet mixture models. *Evolving Systems*, 3(3):153–165, Sep 2012.
26. Taoufik Bdiri and Nizar Bouguila. Positive vectors clustering using inverted Dirichlet finite mixture models. *Expert Systems with Applications*, 39(2):1869–1882, 2012.
27. Huan Liu and Rudy Setiono. Some issues on scalable feature selection1this is an extended version of the paper presented at the fourth world congress of expert systems: Application of advanced information technologies held in Mexico City in March 1998.1. *Expert Systems with Applications*, 15(3):333–339, 1998.

28. Jing Zhou, Dean P. Foster, Robert A. Stine, and Lyle H. Ungar. Streamwise feature selection. *J. Mach. Learn. Res.*, 7:1861–1885, December 2006.
29. Lyle H. Ungar, Jing Zhou, Dean P. Foster, and Bob A. Stine. Streaming feature selection using iic. In Robert G. Cowell and Zoubin Ghahramani, editors, *aistats05*, pages 357–364. Society for Artificial Intelligence and Statistics, 2005.
30. David Dilts, Joseph Khamalah, and Ann Plotkin. Using cluster analysis for medical resource decision making. *Medical Decision Making*, 15(4):333–346, 1995.
31. GJ McLachlan. Cluster analysis and related techniques in medical research. *Statistical Methods in Medical Research*, 1(1):27–48, 1992.
32. Charles Romesburg and Kim Marshall. User's manual for cluster/clustid computer programs for hierarchical cluster analysis. 1984.
33. Jane Clatworthy, Deanna Buick, Matthew Hankins, John Weinman, and Robert Horne. The use and reporting of cluster analysis in health psychology: A review. *British journal of health psychology*, 10(3):329–358, 2005.
34. Matthew R Weir, Edward W Maibach, George L Bakris, Henry R Black, Purnima Chawla, Franz H Messerli, Joel M Neutel, and Michael A Weber. Implications of a health lifestyle and medication analysis for improving hypertension control. *Archives of internal medicine*, 160(4):481–490, 2000.
35. Minlei Liao, Yunfeng Li, Farid Kianifard, Engels Obi, and Stephen Arcona. Cluster analysis and its application to healthcare claims data: a study of end-stage renal disease patients who initiated hemodialysis. *BMC nephrology*, 17(1):25, 2016.
36. Michael B Eisen, Paul T Spellman, Patrick O Brown, and David Botstein. Cluster analysis and display of genome-wide expression patterns. *Proceedings of the National Academy of Sciences*, 95(25):14863–14868, 1998.
37. Eman Abdel-Maksoud, Mohammed Elmogy, and Rashid Al-Awadi. Brain tumor segmentation based on a hybrid clustering technique. *Egyptian Informatics Journal*, 16(1):71–81, 2015.
38. Youyong Kong, Yue Deng, and Qionghai Dai. Discriminative clustering and feature selection for brain mri segmentation. *IEEE Signal Processing Letters*, 22(5):573–577, 2014.
39. Jianhua Yao, Jeremy Chen, and Catherine Chow. Breast tumor analysis in dynamic contrast enhanced mri using texture features and wavelet transform. *IEEE Journal of selected topics in signal processing*, 3(1):94–100, 2009.
40. Paul Juneau. Analyzing pregnancy costs with finite mixture models: An opportunity to more adequately accommodate the presence of patient data heterogeneity. *Gynecology and Obstetrics Research - Open Journal*, 2:69–76, 09 2015.
41. Mohamed Al Mashrgy, Taoufik Bdiri, and Nizar Bouguila. Robust simultaneous positive data clustering and unsupervised feature selection using generalized inverted Dirichlet mixture models. *Knowledge-Based Systems*, 59:182–195, 2014.
42. Geoffrey McLachlan and David Peel. *Finite mixture models*. John Wiley & Sons, 2004.
43. Zhanyu Ma and Arne Leijon. Bayesian estimation of beta mixture models with variational inference. *IEEE Transactions on Pattern Analysis and Machine Intelligence*, 33(11):2160–2173, 2011.
44. Christopher Bishop and John Winn. Structured variational distributions in vibes. 2003.
45. D. Chandler. *Introduction to Modern Statistical Mechanics*. September 1987.
46. Gilles Celeux, Florence Forbes, and Nathalie Peyrard. Em procedures using mean field-like approximations for Markov model-based image segmentation. *Pattern recognition*, 36(1):131–144, 2003.
47. Adrian Corduneanu and Christopher M Bishop. Variational Bayesian model selection for mixture distributions. In *Artificial intelligence and Statistics*, volume 2001, pages 27–34. Morgan Kaufmann Waltham, MA, 2001.
48. Masa-Aki Sato. Online model selection based on the variational Bayes. *Neural computation*, 13(7):1649–1681, 2001.
49. Shun-Ichi Amari. Natural gradient works efficiently in learning. *Neural computation*, 10(2):251–276, 1998.

50. Matthew Hoffman, Francis R Bach, and David M Blei. Online learning for latent Dirichlet allocation. In *advances in neural information processing systems*, pages 856–864, 2010.
51. Harold Kushner and G George Yin. *Stochastic approximation and recursive algorithms and applications*, volume 35. Springer Science & Business Media, 2003.
52. Ayşe Demirhan, Mustafa Törü, and Inan Güler. Segmentation of tumor and edema along with healthy tissues of brain using wavelets and neural networks. *IEEE journal of biomedical and health informatics*, 19(4):1451–1458, 2014.
53. K Roy and J Mukherjee. Image similarity measure using color histogram, color coherence vector, and Sobel method. *International Journal of Science and Research*, 2:538–543, 01 2013.
54. Komal Sharma, Akwinder Kaur, and Shruti Gujral. Brain tumor detection based on machine learning algorithms. *International Journal of Computer Applications*, 103(1), 2014.
55. Nilesh Bhaskarrao Bahadure, Arun Kumar Ray, and Har Pal Thethi. Image analysis for mri based brain tumor detection and feature extraction using biologically inspired bwt and svm. *International journal of biomedical imaging*, 2017, 2017.
56. Samir Kumar Bandhyopadhyay and Tuhin Utsab Paul. Automatic segmentation of brain tumour from multiple images of brain mri. *Int J Appl Innovat Eng Manage (IJAIEM)*, 2(1):240–8, 2013.
57. A. Meena and R. Raja. Spatial fuzzy c means pet image segmentation of neurodegenerative disorder. *ArXiv*, abs/1303.0647, 2013.
58. Cosmin Cernazanu-Glavan and Stefan Holban. Segmentation of bone structure in x-ray images using convolutional neural network. *Adv. Electr. Comput. Eng*, 13(1):87–94, 2013.
59. Amit Yerpude and Sipi Dubey. Colour image segmentation using k-medoids clustering. *Int J Comput Technol Appl*, 3(1):152–4, 2012.
60. Md. Mostafa Kamal Sarker, Hatem A. Rashwan, Mohamed Abdel-Nasser, Vivek Kumar Singh, Syeda Furruka Banu, Farhan Akram, Forhad U H Chowdhury, Kabir Ahmed Choudhury, Sylvie Chambon, Petia Radeva, and Domenec Puig. Mobilegan: Skin lesion segmentation using a lightweight generative adversarial network, 2019.
61. Dev Kumar Das, Madhumala Ghosh, Mallika Pal, Asok K Maiti, and Chandan Chakraborty. Machine learning approach for automated screening of malaria parasite using light microscopic images. *Micron*, 45:97–106, 2013.
62. C. Constantinopoulos and A. Likas. Unsupervised learning of Gaussian mixtures based on variational component splitting. *IEEE Transactions on Neural Networks*, 18(3):745–755, May 2007.

Chapter 9
Entropy-Based Variational Inference for Semi-Bounded Data Clustering in Medical Applications

Narges Manouchehri, Maryam Rahmanpour, and Nizar Bouguila

Abstract Over the past decades, the unprecedented availability of various types of data and simultaneous development of technology established extensive interest in applying numerous machine learning approaches to extract the implicit patterns, acquire information and retrieve latent meaningful knowledge. Such powerful statistical tools have been applied in various fields of science.

One of the vital domains where these techniques could be potentially deployed is healthcare. The main intention in this field is that medical diagnosis procedures and healthcare examinations generate a huge amount of various data types such as text, image, video and signal. Dealing with such large complex data is beyond the scope of human competence. Consequently, machine learning tools are significantly valuable as they assist the clinicians in processing medical datasets, achieving broader insight, planning and managing diseases, providing better care which leads to having better outcomes including elimination of unnecessary costs and increasing patient satisfaction.

In this work, we focus on one of the main clustering methods of machine learning approaches, namely mixture models. These capable techniques have demonstrated high potential and flexibility to express data. Gaussian mixture models (GMM) have been widely applied in various fields of research to express symmetric data. However, for asymmetric and non-Gaussian data, other alternatives such as inverted Dirichlet mixture models could describe the data more accurately. To learn our model, we employ an entropy-based variational approach and then evaluate it on four medical applications.

N. Manouchehri (✉) · N. Bouguila · M. Rahmanpour
Concordia Institute for Information Systems Engineering (CIISE), Concordia University, Montreal, QC, Canada
e-mail: narges.manouchehri@mail.concordia.ca; nizar.bouguila@concordia.ca

M. Masmoudi et al. (eds.), *Artificial Intelligence and Data Mining in Healthcare*,
https://doi.org/10.1007/978-3-030-45240-7_9

9.1 Introduction

The enthusiasm about machine learning (ML) models is augmenting markedly due to their substantial potential as powerful inference engines for modelling heterogeneous and multimodal data [1] and automatic extraction of knowledge. In this context, among numerous emerged statistical approaches, clustering algorithms as unsupervised methods grabbed lots of attention due to their ability to fulfil more complex tasks and deal with the unlabelled data without needing any manual intervention. As one of their main frameworks, finite mixture models have been widely adopted due to their flexibility and simplicity [2]. In this modelling method, we assume that the data includes a finite number of latent clusters, each described by a statistical distribution.

When dealing with these probabilistic efficient techniques, we face three major challenges. Firstly, selection of the appropriate probability density function which well describes the data. Secondly, estimation of the model's parameter and lastly defining the proper number of clusters or determination of model complexity. To tackle the first issue, Gaussian mixture models (GMM) have been commonly applied in classical researches assuming that the data has symmetric nature [3]. Nevertheless, in real-world application, the data has non-Gaussian asymmetric characteristics. Thus, other alternatives such as Dirichlet [4], inverted Dirichlet [5–7], generalized Dirichlet [8], Beta-Liouville [9] and inverted Beta-Liouville [10] mixture models have been recently used .

To learn finite mixture models and estimate their parameters, several approaches have been extensively studied and two main branches of techniques have been proposed, namely frequentist [11] and Bayesian [12]. Each of these has some drawbacks. For instance, maximum likelihood as a deterministic algorithm suffers from dependency on initialization [13], overfitting and convergence to local maxima [14] instead of the global maximum. To overcome such problems, pure Bayesian techniques such as Markov chain Monte Carlo (MCMC) have been suggested. However, their high computational costs and difficult assessment in the inference of convergence are their major drawbacks [15, 16]. Thus, to beneficiate from the advantages of a deterministic and fully Bayesian approaches and avoid their disadvantages, variational Bayes approaches were proposed [17–20] as they are more controllable, have less computational burden compared to MCMC techniques and tackle the problem of overfitting [21–23]. The main concept is to approximate the model posterior distribution. To do that, we minimize the Kullback–Leibler (KL) divergence between the true posterior and an approximating distribution [24, 25].

Estimating model complexity is the last challenge when dealing with mixture models. Several approaches such as minimum message length criterion [26] were studied previously. In other researches, some splitting approaches have been proposed [17] for determining the number of mixture clusters. For instance, an entropy-based variational algorithm has been developed to learn Gaussian mixture models [27]. In this case, the theoretical and estimated entropies are compared to split the component. This approach was previously studied for finite Dirichlet [28] and Beta-Liouville [29] mixture models.

The objective of our work is to develop a variational framework to learn finite inverted Dirichlet mixture models via entropy-based component splitting approach to perform parameters estimation and model selection simultaneously. We validated our proposed frameworks through real-world medical applications. In recent years, machine learning gained lots of attention from researchers and became influential in healthcare. Indeed, considering impressing advances in medical equipment technologies and digitizing health records, an incredible amount of clinical information of various types are generated. Thus, collection, effective analysis and accurate interpretation of medical test results are great challenges. Moreover, erroneous and imprecise inference impose unnecessary costs to the healthcare system. More importantly, such diagnostic errors may inflict severe harms and irreversible consequences to patients due to delayed disease management. Thus, implementations of machine learning architectures for processing and handling large complex datasets promise risk stratification and patients satisfaction in various fields of medicine [30].

In this book, another chapter is assigned to "online variational learning using finite generalized inverted Dirichlet mixture model with feature selection on medical datasets". The differences between the above-mentioned chapter and our work are as follows:

- The distribution in our work is inverted Dirichlet, while the mixture model in the above-mentioned chapter is developed based on generalized inverted Dirichlet distribution.
- In entropy-based variational inference, we focus on model selection and splitting approach with the help of entropy to tackle the problem of finding the proper number of clusters, while online variational method focuses on the sequential process of data instances which is important specifically in modelling large-scale data and real-time applications. Online models are considerably faster than batch variational technique.
- Our applications are related to applying machine learning techniques in medical data analysis, whereas in the other chapter the authors focused on medical applications associated with computer vision including image segmentation and image classification.

The remainder of this paper is organized as follows: Section 9.2 presents the details of our finite mixture model. In Sect. 9.3, we describe entropy-based variational approximation procedure for the proposed model learning. Section 9.4 is devoted to experimental studies considering four medical real-world applications including cardiovascular diseases prediction, diabetes detection, lung and breast cancer prediction. Finally, Sect. 9.5 concludes the paper and gives some perspectives.

9.2 Finite Inverted Dirichlet Mixture Model

If a D-dimensional positive vector $\mathbf{X}_i = (x_{i1}, \ldots, x_{iD})$ is sampled from a finite inverted Dirichlet ($\mathcal{ID}$) mixture model with M components, then we have

$$p\left(\mathbf{X_i} \mid \boldsymbol{\pi}, \boldsymbol{\alpha}\right) = \sum_{j=1}^{M} \pi_j \mathcal{ID}\left(\mathbf{X}_i \mid \boldsymbol{\alpha}_j\right) \tag{9.1}$$

where $\boldsymbol{\alpha} = (\boldsymbol{\alpha}_1, \ldots, \boldsymbol{\alpha}_M)$ and $\boldsymbol{\pi} = (\pi_1, \ldots, \pi_M)$ denote the mixing coefficients with the constraints that they are positive and sum to one. $\mathcal{ID}\left(\mathbf{X}_i \mid \boldsymbol{\alpha}_j\right)$ represents the jth inverted Dirichlet distribution with parameter $\boldsymbol{\alpha}_j$ and is defined as [31]:

$$\mathcal{ID}\left(\mathbf{X}_i \mid \boldsymbol{\alpha}_j\right) = \frac{\Gamma\left(\sum_{l=1}^{D+1} \alpha_{jl}\right)}{\prod_{l=1}^{D+1} \Gamma\left(\alpha_{jl}\right)} \prod_{l=1}^{D} x_{il}^{\alpha_{jl}-1} \left(1 + \sum_{l=1}^{D} x_{il}\right)^{-\sum_{l=1}^{D+1} \alpha_{jl}} \tag{9.2}$$

where $0 < x_{il} < \infty$ for $l = 1, \ldots, D$. In addition, $\boldsymbol{\alpha}_j = \left(\alpha_{j1}, \alpha_{j2}, \ldots, \alpha_{jD}\right)$ such that $\alpha_{jl} > 0$ for $l = 1, \ldots, D+1$.

Next, we introduce an M-dimensional binary random vector $\mathbf{Z_i} = \{Z_{il}, \ldots, Z_{iM}\}$ for each observed vector $\mathbf{X_i}$, such that $Z_{ij} \in \{0, 1\}, \sum_{j=1}^{M} Z_{ij} = 1$ and $Z_{ij} = 1$ if $\mathbf{X_i}$ belongs to component j and 0, otherwise. Notice that, $\mathcal{Z} = \{\mathbf{Z}_1, \ldots, \mathbf{Z}_N\}$ are called the membership vectors of the mixture model and also considered as the latent variables since they are actually hidden variables that do not appear explicitly in the model. Furthermore, the conditional distribution of $\mathcal{Z}$ given the mixing coefficients $\boldsymbol{\pi}$ is defined as:

$$p\left(\mathcal{Z} \mid \boldsymbol{\pi}\right) = \prod_{j=1}^{N} \prod_{j=1}^{M} \pi_j^{Z_{ij}} \tag{9.3}$$

Then, the likelihood function with latent variables which is indeed the conditional distribution of dataset $\mathcal{X}$ given the class labels $\mathcal{Z}$ can be written as:

$$p\left(\mathcal{X} \mid \mathcal{Z}, \boldsymbol{\alpha}\right) = \prod_{j=1}^{N} \prod_{j=1}^{M} \mathcal{ID}\left(\mathbf{X}_i \mid \boldsymbol{\alpha_j}\right)^{Z_{ij}} \tag{9.4}$$

Moreover, we assume that the parameters of the inverted Dirichlet are statistically independent and for each parameter α_{jl}, the Gamma distribution $\mathcal{G}$ is adopted to approximate the conjugate prior:

$$p\left(\alpha_{jl}\right) = \mathcal{G}\left(\alpha_{jl} \mid u_{jl}, v_{jl}\right) = \frac{v_{jl}^{u_{jl}}}{\Gamma\left(u_{jl}\right)} \alpha_{jl}^{u_{jl}^{-1}} e^{-v_{jl}\alpha_{jl}} \tag{9.5}$$

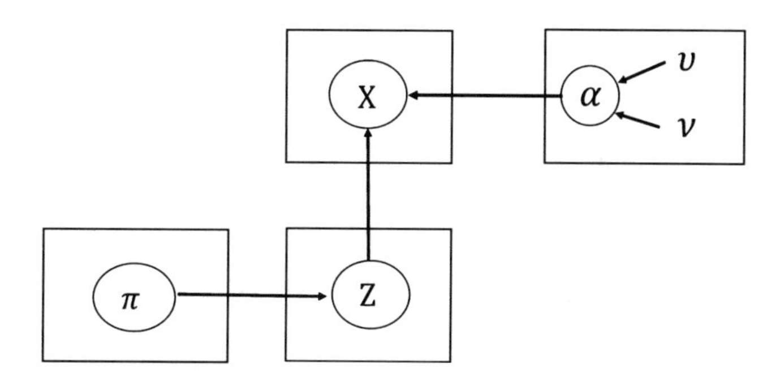

Fig. 9.1 Graphical model of the finite inverted Dirichlet mixture. Symbols in circles denote random variables; otherwise, they denote model parameters. The conditional dependencies of the variables are represented by the arcs

where u_{jl} and υ_{jl} are positive hyperparameters.

Thus, the joint distribution of all the random variables, conditioned on the mixing coefficients can be written as:

$$
\begin{aligned}
p(\mathcal{X}, \mathcal{Z}, \boldsymbol{\alpha} \mid \boldsymbol{\pi}) &= p(\mathcal{X} \mid \mathcal{Z}, \boldsymbol{\alpha})\, p(\mathcal{Z} \mid \boldsymbol{\pi})\, p(\boldsymbol{\alpha}) \\
&= \prod_{i=1}^{N} \prod_{j=1}^{M} \left[\pi_j \frac{\Gamma\left(\sum_{l=1}^{D+1} \alpha_{jl}\right)}{\prod_{l=1}^{D+1} \Gamma\left(\alpha_{jl}\right)} \prod_{l=1}^{D} x_{il}^{\alpha_{jl}-1} \times \left(1 + \sum_{l=1}^{D} x_{il}\right)^{-\sum_{l=1}^{D+1} \alpha_{jl}} \right]^{Z_{ij}} \\
&\times \prod_{j=1}^{M} \prod_{l=1}^{D+1} \frac{\upsilon_{jl}^{u_{jl}}}{\Gamma\left(u_{jl}\right)} \alpha_{jl}^{u_{jl}-1} e^{-\upsilon_{jl}\alpha_{jl}}
\end{aligned}
\tag{9.6}
$$

A graphical representation of this model is shown in Fig. 9.1.

9.3 Entropy-Based Variational Learning

9.3.1 Variational Learning

Variational Bayes inference is used to approximate the posterior distribution $p(\Theta \mid \mathcal{X})$, where $\mathcal{X}$ is a set of observed variables and Θ represents the set of all latent variables and parameters. Considering the Kullback–Leibler (KL) divergence between the approximating $Q(\Theta)$ and the true posterior distribution $p(\Theta \mid \mathcal{X})$ we have [19]

$$KL(Q \mid\mid P) = -\int Q(\Theta) \ln\left(\frac{p(\Theta \mid \mathcal{X}, \boldsymbol{\pi})}{Q(\Theta)}\right) d\Theta = \ln p(\mathcal{X} \mid \boldsymbol{\pi}) - \mathcal{L}(Q) \tag{9.7}$$

$$\mathcal{L}(Q) = \int Q(\Theta) \ln\left(\frac{p(\mathcal{X}, \Theta \mid \boldsymbol{\pi})}{Q(\Theta)}\right) d\Theta \tag{9.8}$$

Since the KL divergence satisfies $KL(Q \mid\mid p) \geq 0$, it follows that $\mathcal{L}(Q) \leq \ln p(\mathcal{X})$, which means that $\mathcal{L}(Q)$ represents the lower bound on $\ln p(\mathcal{X})$. Then, maximizing the lower bound $\mathcal{L}(Q)$, we can obtain the true posterior distribution which is equivalent to minimizing the KL divergence. Obviously, when $Q(\Theta) = p(\Theta \mid \mathcal{X})$ the KL divergence is minimized and equals to zero. In practice, working with the true posterior distribution is often intractable. Thus, in our study mean-field is adopted. As a result, $Q(\Theta)$ is considered as the factorization of disjoint tractable distributions as follows [1, 21, 32]:

$$Q(\Theta) = Q(\mathcal{Z})Q(\boldsymbol{\alpha}) \tag{9.9}$$

We use variational optimization to maximize the lower bound $\mathcal{L}(Q)$. The general expression for its optimal solution is as follows:

$$Q_s(\Theta_s) = \frac{\exp \langle \ln\ p(\mathcal{X}, \Theta) \rangle_{j \neq s}}{\int \exp \langle \ln\ p(\mathcal{X}, \Theta) \rangle_{j \neq s} d\Theta} \tag{9.10}$$

where $\exp(\cdot)$ is exponential function and $\langle . \rangle_{s \neq j}$ represents an expectation with respect to all variational factors except for the s-th one.
We can obtain the following variational solutions for the finite inverted Dirichlet mixture model [5]:

$$Q(\mathcal{Z}) = \prod_{j=1}^{M} r_{ij}^{Z_{ij}} \tag{9.11}$$

$$Q(\boldsymbol{\alpha}) = \prod_{j=1}^{M} \prod_{d=1}^{D} \mathcal{G}\left(\alpha_{jd} \mid u_{jd}^{*}, v_{jd}^{*}\right) \tag{9.12}$$

$$r_{ij} = \frac{\rho_{ij}}{\sum_{j=1}^{M} \rho_{ij}} \tag{9.13}$$

$$\rho_{ij} = exp\left\{ \ln \pi_j + \tilde{R}_j + \sum_{l=1}^{D} (\bar{\alpha}_{jl} - 1) \ln x_{il} - \left(\sum_{l=1}^{D+1} \bar{\alpha}_{jl}\right) \ln\left(1 + \sum_{l=1}^{D} x_{il}\right) \right\} \tag{9.14}$$

$$\tilde{R}_j = \ln \frac{\Gamma\left(\sum_{l=1}^{D+1} \overline{\alpha}_{jl}\right)}{\prod_{l=1}^{D+1} \Gamma\left(\overline{\alpha}_{jl}\right)} + \sum_{l=1}^{D+1} \overline{\alpha}_{jl} \left[\psi\left(\sum_{l=1}^{D+1} \overline{\alpha}_{jl}\right) - \psi\left(\overline{\alpha}_{jl}\right)\right]$$

$$\times \left[\langle \ln \alpha_{jl} \rangle - \ln \overline{\alpha}_{jl}\right] + \frac{1}{2} \sum_{l=1}^{D+1} \overline{\alpha}_{jl}^2 \left[\psi'\left(\sum_{l=1}^{D+1} \overline{\alpha}_{jl}\right) - \psi'\left(\overline{\alpha}_{jl}\right)\right]$$

$$\times \left\langle \left(\ln \alpha_{jl} - \ln \overline{\alpha}_{jl}\right)^2 \right\rangle + \frac{1}{2} \sum_{a=1}^{D+1} \sum_{b=1, a \neq b}^{D} \overline{\alpha}_{ja} \overline{\alpha}_{jb} \left\{ \psi'\left(\sum_{l=1}^{D+1} \overline{\alpha}_{jl}\right)\right.$$

$$\left. \times \left(\langle \ln \overline{\alpha}_{ja} \rangle - \ln \overline{\alpha}_{ja}\right) \times \left(\langle \ln \overline{\alpha}_{jb} \rangle - \ln \overline{\alpha}_{jb}\right) \right\} \tag{9.15}$$

$$u_{jl}^* = u_{jl} + \sum_{i=1}^{N} \langle Z_{ij} \rangle \bar{\alpha}_{jl} \left[\psi\left(\sum_{l=1}^{D+1} \bar{\alpha}_{jl}\right) - \psi\left(\bar{\alpha}_{jl}\right)\right.$$

$$\left. + \sum_{k \neq l}^{D+1} \bar{\alpha}_{jk} \psi'\left(\sum_{l=1}^{D+1} \bar{\alpha}_{jl}\right) \left(\langle \ln \alpha_{jk} \rangle - \ln \bar{\alpha}_{jk}\right)\right] \tag{9.16}$$

$$v_{jl}^* = v_{jl} - \sum_{i=1}^{N} \langle Z_{ij} \rangle \left[\ln x_{il} - \ln\left(1 + \sum_{l=1}^{D} x_{il}\right)\right] \tag{9.17}$$

where $\psi(.)$ is digamma function.
The expected values in the above formulas are

$$\langle Z_{ij} \rangle = r_{ij} \tag{9.18}$$

$$\bar{\alpha}_{jl} = \langle \alpha_{jl} \rangle = \frac{u_{jl}}{v_{jl}} \tag{9.19}$$

$$\langle \ln \alpha_{jl} \rangle = \psi\left(u_{jl}\right) - \ln v_{jl} \tag{9.20}$$

Note that R_j is defined as:

$$R_j = \left\langle \ln \frac{\Gamma\left(\sum_{l=1}^{D+1} \alpha_{jl}\right)}{\prod_{l=1}^{D+1} \Gamma\left(\alpha_{jl}\right)} \right\rangle \tag{9.21}$$

$\tilde{R}_j$ is the approximate lower bound of it. Since a closed form expression cannot be found for R_j, the standard variational inference cannot be applied directly and second-order Taylor series expansion is applied to find a lower bound approximation [33]. In our case, point estimations of the values of π_j are evaluated by maximizing the variational likelihood bound $\mathcal{L}(Q)$. Setting the derivative of this lower bound with respect to π_j to zero gives

$$\pi_j = \frac{1}{N} \sum_{i=1}^{N} r_{ij} \tag{9.22}$$

9.3.2 Model Learning Through Entropy-Based Variational Bayes

Mainly motivated by [27], we develop an entropy-based variational Bayes for learning the inverted Dirichlet mixture model. The fundamental idea is that we can test if a given component was truly inverted Dirichlet distributed by comparing its theoretical maximum entropy with the one estimated by the MeanNN estimator [34]. Significant difference from this comparison represents that this component is not well fitted, it is then split into two new components with a proper initialization.

9.3.3 Theoretical Entropy of Inverted Dirichlet Mixtures

Differential entropy is the entropy measure of continuous probability distributions. For a continuous random variable $\mathbf{X}$ with N possible values $\{\mathbf{X}_1, \ldots, \mathbf{X}_N\}$ and probability density function $p(\mathbf{X})$, the differential entropy is calculated by [28, 29]

$$H\left(\mathbf{X}_i\right) = -\int p\left(\mathbf{X}_i\right) \log_2 p\left(\mathbf{X}_i\right) d\mathbf{X}_i \tag{9.23}$$

The maximum differential entropy in the case of an inverted Dirichlet distribution is given by

$$H_{ID}\left[p\left(\mathbf{X}_i \mid \theta\right)\right] = -\log \Gamma\left(\sum_{l=1}^{D+1} \alpha_l\right) + \sum_{l=1}^{D+1} \log \Gamma\left(\alpha_l\right)$$
$$-\sum_{l=1}^{D} \left(\alpha_l - 1\right)\left(\psi\left(\alpha_l\right) - \psi\left(\sum_{l=1}^{D+1} \alpha\right)\right) + \left(\sum_{l=1}^{D+1} \alpha_l\right) \psi\left(\sum_{l=1}^{D+1} \alpha_l\right) \tag{9.24}$$

9.3.4 MeanNN Entropy Estimator

In order to test if a given component is truly distributed according to inverted Dirichlet distribution, we adopt the MeanNN entropy estimator[34]. The aim of MeanNN entropy estimator is to estimate $H(\mathbf{X})$ of a D-dimensional random variable $\mathbf{X}$ with an unknown density function $p(\mathbf{X})$[35]. Shannon differential entropy can be considered as the average of $-\log p(\mathbf{X})$ and an unbiased entropy estimator can be formed by estimating $\log p(\mathbf{X})$.

Suppose that a ball with diameter ϵ is centred at $\mathbf{X}_i$ and there exists a point within distance $[\epsilon, \epsilon + d_\epsilon]$ from $\mathbf{X}_i$, then there are $\hat{k} - 1$ other points at a smaller distances and $N - \hat{k} - 1$ points have larger distances from $\mathbf{X}_i$. According to this assumption, the probability of the distance between $\mathbf{X}_i$ and its $\hat{k}$-th nearest neighbour can be obtained by [34]

$$P_{i\hat{k}}(\epsilon) = \frac{(N-1)!}{\left(\hat{k}-1\right)!\left(N-\hat{k}-1\right)!} \frac{dp_i(\epsilon)}{d\epsilon} p_i^{\hat{k}-1} (1-p_i)^{N-\hat{k}-1} \tag{9.25}$$

where $p_i(\epsilon)$ represents the mass of the ϵ-ball centred at $\mathbf{X}_i$, and:

$$p_i(\epsilon) = \int_{||\mathbf{X}-\mathbf{X}_i||<\epsilon} p(\mathbf{X}) d\mathbf{X} \tag{9.26}$$

So, the expectation of $\log p_i(\epsilon)$ with respect to $p_i(\epsilon)$ is given by

$$E(\log\ p_i(\epsilon)) = \int_0^\infty P_{i\hat{k}} \log\ p_i(\epsilon)\, d\epsilon = \psi\left(\hat{k}\right) - \psi(N) \tag{9.27}$$

Suppose that $p(\mathbf{X})$ is constant in the entire ϵ-ball and is equal to:

$$p_i(\epsilon) \simeq V_d \epsilon^d p(\mathbf{X}_i) \tag{9.28}$$

where d represents the dimension of $\mathbf{X}$, and V_d denotes the volume of the unit ball which is calculated by

$$V_d = \pi^{d/2} / \Gamma(1 + d/2) \tag{9.29}$$

By substituting $p_i(\epsilon)$ into (9.27) and applying the unbiased kNN estimator, the differential entropy can be calculated by

$$H_{\hat{k}}(\mathbf{X}) = \psi(N) - \psi\left(\hat{k}\right) + \frac{d}{N} \sum_{i=1}^{N} lod\epsilon_i + \log V_d \tag{9.30}$$

By considering all possible values of $\hat{k}$ from 1 to $N-1$, the MeanNN estimator of the differential entropy is

$$H_M(\mathbf{X}) = \frac{1}{N-1}\sum_{\hat{k}=1}^{N-1} H_{\hat{k}}(\mathbf{X}) \tag{9.31}$$

$$= \log V_d + \psi(N) + \frac{1}{N-1}\sum_{\hat{k}=1}^{N-1}\left[\frac{d}{N}\sum_{i=1}^{N}\log \epsilon_{i,\hat{k}} - \psi\left(\hat{k}\right)\right]$$

where $\epsilon_{i,\hat{k}}$ denotes the $\hat{k}$-th nearest neighbour of $\mathbf{X}_i$.
The maximum entropy of the inverted Dirichlet mixture model can be obtained by

$$H_{ID} = \sum_{j=1}^{M} \pi_j H_{ID}(j) \tag{9.32}$$

Similar to [27], we define Ω_{ID} as the normalized weighted sum of the difference between the theoretical and the estimated entropy of each component associated with the inverted Dirichlet mixture model as:

$$\Omega_{ID} = \sum_{j=1}^{M}\left[\frac{H_{ID}(j) - H_M(j)}{H_{ID}(j)}\right] = \sum_{j=1}^{M} \pi_j \left[1 - \frac{H_M(j)}{H_{ID}(j)}\right] \tag{9.33}$$

As a result, Ω_{ID} is defined in [0,1], and $\Omega_{ID} = 0$ only if data was truly inverted Dirichlet distributed.
In the splitting process, we choose the component j^* with the highest $\Omega_{ID}(j)$ as:

$$j^* = \arg\max_j [\Omega_{ID}(j)] = \arg\max_j \left[\pi_j \frac{H_{ID}(j) - H_M(j)}{H_{ID}(j)}\right] \tag{9.34}$$

and then we replace j^* by two new components.
In order to use the variational Bayes for learning inverted Dirichlet mixture models, we begin with only one component ($M = 1$) with properly initialized parameters. The algorithm is summarized below.

9.4 Experimental Results

To validate the inverted Dirichlet mixture model with entropy-based variational learning, four real medical datasets have been put into work. We focus on analysing four main causes of mortality, namely cardiovascular diseases, diabetes, breast and lung cancer. With the help of the confusion matrix, we compared the accuracy of our

Algorithm 1: Entropy-based variational learning algorithm

1. Initialization:
   - $M = 1$, $j^* = M$, $\pi_1 = 1$.
   - Initialize hyperparameters u_{jl}, v_{jl}.
2. The splitting process:
   - Split j^* into two new components j_1 and j_2 with equal proportion $\pi^*/2$.
   - $M = M + 1$.
   - Initialize the parameters of j_1 and j_2 using same parameters of j^*.
3. Apply standard variational Bayes, until convergence.
4. Determine the number of components through the evaluation of the mixing coefficients $\{\pi_j\}$ according to (9.22).
5. If $\pi_j \simeq 0$, where $j \in 1, \ldots, M$ then $M = M - 1$ and program terminates.
6. else evaluate Ω_{MD}, choose j^* according to (9.34), and go to the splitting process in step 2.

model performance, entropy-based variational learning of inverted Dirichlet (EV-IDMM) with entropy-based variational inference of Dirichlet (EV-DMM) and batch variational inference of three other mixture models based on Dirichlet, inverted Dirichlet and Gaussian distributions which are denoted here by BV-DMM, BV-IDMM and BV-GMM, respectively.

9.4.1 Cardiovascular Diseases (CVDs)

Cardiovascular diseases (CVDs) as a vast variety of disorders affecting the heart and blood vessels are recognized as the first ranked cause of global death. This leading reason of mortality has taken the lives of 17.9 million people in 2016 [36]. Diagnosis and monitoring of these diseases require highly expensive equipment and according to the necessity of long term treatment, lots of costs are imposed on the healthcare system. Thus, the side effects of the CVDs damage personal life of patients by reduction of productivity and simultaneously lead to socio-economic harms and cost burdens. However, in the majority of cases, the associated risk factors of these diseases such as tobacco use, obesity, low level of physical activity and diet could be addressed. As a consequence, prevention has a significant role in

Table 9.1 Accuracy of model performance in CVDs classification

Method	Accuracy (%)
EV-IDMM	**91.33**
EV-DMM	89.2
BV-IDMM	86.9
BV-DMM	88.1
BV-GMM	83.6

The best result is shown in bold

CVDs management. To achieve this goal, early detection, making a proper decision and taking necessary actions are essential.

Nowadays, physicians analyse patient clinical history, biomarkers, test results including images, signals and text. The inference based on such complex data depends on their expertise in addition to other factors such as time and workload. Thus, such diagnosis procedure could be error-prone, inefficient and imprecise which could result in decreasing the quality of healthcare service, threatening the lives and waste of resources. In this situation, the asset of automation in medical inference is enhanced [37].

In our work, we applied our proposed algorithm on a real dataset [38] to predict the presence of heart disease. This dataset contains 303 instances considering 13 attributes including age, sex, chest pain type (typical and atypical angina or non-anginal pain), resting blood pressure, serum cholesterol level, fasting blood sugar, resting electrocardiographic results (normal, ST-T wave abnormality or left ventricular hypertrophy), maximum heart rate achieved, exercise induced angina, ST depression peak, the slope of the peak exercise ST segment (upsloping, flat or downsloping), number of major vessels coloured by fluoroscopy and type of defect (normal, fixed or reversible). The data is labelled by angiographic disease status distinguishing absence from presence. The result of our evaluation is presented in Table 9.1 indicating the outperformance of EV-IDMM.

9.4.2 *Diabetes*

Diabetes is an incurable but controllable metabolic and chronic illness characterized by hyperglycaemia, affecting the 422 million patients in 2014. 1.6 million deaths were reported due to diabetes as the seventh leading cause of death in 2016 [39]. If untreated, unfortunately this disease affects several vital organs by developing lots of diabetes-related complications such as blindness, nerve damage, kidney failure, heart attacks, stroke and lower limb amputation [39]. Moreover, the prevalence of diabetes has been increasing speedily and alarmingly. By following a healthy lifestyle such as maintaining the weight in the normal range, avoiding tobacco consumption, having healthy eating habits and frequent physical activity, it is possible to avoid, delay or manage diabetes, its associated side effects and irreversible complications.

Table 9.2 Accuracy of model performance in diabetes detection

Method	Accuracy (%)
EV-IDMM	**94.06**
EV-DMM	90.29
BV-IDMM	93.5
BV-DMM	85.4
BV-GMM	87.06

Therefore, how to rapidly and precisely diagnose and manage this disease is a topic of worthy researches. The earlier diagnosis is obtained, the much easier it is handled. This warning figure clarifies the distinction of applying statistical tools to process, analyse patient data and predict hazards in advance. Thus, numerous machine learning algorithms are helping physicians to predict diabetes [40].

To validate our framework, we examined it on a dataset [41] to predict the presence or absence of diabetes. This dataset includes 268 positive and 500 negative cases which were described by plasma glucose concentration, blood pressure, diastolic blood pressure (mm Hg), Triceps skinfold thickness (mm), 2-h serum insulin level (mu U/ml), body mass index (BMI), diabetes pedigree function, number of pregnancies and age. Table 9.2 indicates the robustness of our proposed model compared to other similar methods.

9.4.3 Lung Cancer

Cancer is the second principal mortality reason in the world and was in charge of approximately 9.6 million deaths in 2018 [42]. Globally, about 1 in 6 deaths is due to cancer. Lung cancer is the most common case with around 2.9 million reported cases and ranked as the first cancer death contributor [42] with 1.76 million deaths records. The incidence and mortality rates of this disease are close and mirror one another which means most of the diagnosed cases eventually ends to death. Similar to the main general risk factors of all cancer types, namely obesity, unhealthy diet, lack of physical activity as well as tobacco and alcohol consumption, these behavioural and dietary risks have a well-connected correlation with lung cancer. Nevertheless, it is noteworthy to mention that among all above-mentioned factors, tobacco use, exposure to second-hand smoke or toxic chemicals have the most distinguished influence. Lung cancers are divided into two major branches, namely non-small and small cell lung cancer which is associated with the cell type from which cancer is raised. The former one which is called adenocarcinoma generally involves glandular cells on the outer part of the lung. The second type, namely, squamous cell carcinoma which appears almost exclusively in heavy smokers, starts in squamous cells which are flat, thin cells lining the bronchi.

Detection of this disease and differentiation between benign and malignant cases is a task of great importance for clinicians. Machine learning (ML) approaches could

Table 9.3 Accuracy of model performance in lung cancer detection

Method	Accuracy (%)
EV-IDMM	**90.41**
EV-DMM	87.22
BV-IDMM	84.09
BV-DMM	80.03
BV-GMM	81.9

be proposed to medical professionals for boosting chances of precise diagnosis and managing incidental of patients at risk and reduce misclassification of cases. One of the most promising applications is discovering patterns in vast amounts of complex data delivered to clinicians and helping them in reaching a new level of understanding and tailoring healthcare services to individual patients.

In this work, we tested the goodness of our model performance on a publicly available dataset [43], containing 32 cases and 56 attributes including 55 features and one class label. As illustrated in Table 9.3, our model works better compared to the others.

9.4.4 *Breast Cancer*

As the most frequent cancer among women and threatening 2.1 million of them yearly, breast cancer is the major cause of women's cancer-related mortality. As it is estimated and reported by WHO [44], 627,000 female patients expired in 2018 due to this crucial disease. To avoid the various damages associated with breast cancer and increase patient survival, it is vital to detect it as much early as possible with the help of frequent check-ups, screening and early diagnosis. This fact enhances the noteworthiness of health system facilities to detect symptoms quickly and be referred to as proper treatment urgently to raise the likelihood of saving the life of patients. To discover abnormal pre-cancerous breast tissues, advanced imaging and pathological equipment are required. Fortunately, such branches of medical technologies had considerable improvements over the past decades. These valuable means assist medical experts to inspect histopathological breast tissue and evaluate cancer aggressiveness considering the sample appearance and morphological characteristics.

In this experiment, we applied our framework on a cytological breast dataset [45] which has 569 sample tissues, 357 benign and 212 malignant cases, including nine cytological characteristics to differentiate between benign and malignant samples. The features include clump thickness (the extent to which epithelial cell aggregates were mono- or multi-layered), marginal adhesion (cohesion of the peripheral cells of the epithelial cell aggregates), the diameter of the population of the largest epithelial cells relative to erythrocytes, bare nuclei (the proportion of single epithelial nuclei that were devoid of surrounding cytoplasm), blandness of nuclear chromatin, normal

Table 9.4 Accuracy of model performance in cytological tissue classification

Method	Accuracy (%)
EV-IDMM	**93.06**
EV-DMM	89.52
BV-IDMM	85.39
BV-DMM	83.77
BV-GMM	82.44

nucleoli, infrequent mitoses, uniformity of epithelial cell size and uniformity of cell shape. The outputs of our evaluation presented in Table 9.4 illustrate that EV-IDMM outperforms the other algorithms.

9.5 Conclusion

The main goal of our study is to learn finite inverted Dirichlet mixture models for modelling and clustering medical data. An entropy-based variational Bayes framework has been proposed for this matter. The parameters of the mixture model are estimated through variational approach. The model selection of this framework is incremental and based on criteria which depict if a given component was truly inverted Dirichlet distributed or not. These criteria are obtained by the difference of theoretical maximum entropy and the one estimated by the MeanNN estimator. The component with the highest difference is chosen to be split into two new components. Through four different medical applications, namely cardiovascular diseases prediction, diabetes detection, lung and breast cancer prediction, the validity of this proposed approach is investigated. It has been shown that our model outperforms other compared models in this paper and leads to promising results for medical applications.

Acknowledgments We express our thanks to the Natural Sciences and Engineering Research Council of Canada (NSERC) as this research was completed by their support.

References

1. Christopher M Bishop. *Pattern recognition and machine learning*. springer, 2006.
2. Geoffrey McLachlan and David Peel. *Finite mixture models*. John Wiley & Sons, 2004.
3. Trevor Hastie and Robert Tibshirani. Discriminant analysis by Gaussian mixtures. *Journal of the Royal Statistical Society: Series B (Methodological)*, 58(1):155–176, 1996.
4. Wentao Fan, Nizar Bouguila, and Djemel Ziou. Variational learning for finite Dirichlet mixture models and applications. *IEEE transactions on neural networks and learning systems*, 23(5):762–774, 2012.

5. Parisa Tirdad, Nizar Bouguila, and Djemel Ziou. Variational learning of finite inverted Dirichlet mixture models and applications. In *Artificial Intelligence Applications in Information and Communication Technologies*, pages 119–145. Springer, 2015.
6. Wentao Fan and Nizar Bouguila. An accelerated variational framework for face expression recognition. In *2018 IEEE International Black Sea Conference on Communications and Networking (BlackSeaCom)*, pages 1–5. IEEE, 2018.
7. Wentao Fan, Can Hu, Jixiang Du, and Nizar Bouguila. A novel model-based approach for medical image segmentation using spatially constrained inverted Dirichlet mixture models. *Neural Processing Letters*, 47(2):619–639, 2018.
8. Nizar Bouguila and Djemel Ziou. A hybrid sem algorithm for high-dimensional unsupervised learning using a finite generalized Dirichlet mixture. *IEEE Transactions on Image Processing*, 15(9):2657–2668, 2006.
9. Nizar Bouguila. Hybrid generative/discriminative approaches for proportional data modeling and classification. *IEEE Transactions on Knowledge and Data Engineering*, 24(12):2184–2202, 2011.
10. Can Hu, Wentao Fan, Ji-Xiang Du, and Nizar Bouguila. A novel statistical approach for clustering positive data based on finite inverted Beta-Liouville mixture models. *Neurocomputing*, 333:110–123, 2019.
11. Geoffrey J McLachlan. Mixture models in statistics. 2015.
12. Tarek Elguebaly and Nizar Bouguila. A hierarchical nonparametric Bayesian approach for medical images and gene expressions classification. *Soft Computing*, 19(1):189–204, 2015.
13. Dimitris Karlis and Evdokia Xekalaki. Choosing initial values for the em algorithm for finite mixtures. *Computational Statistics & Data Analysis*, 41(3–4):577–590, 2003.
14. Kenji Fukumizu and Shun-ichi Amari. Local minima and plateaus in hierarchical structures of multilayer perceptrons. *Neural networks*, 13(3):317–327, 2000.
15. Michael Evans, Tim Swartz, et al. Methods for approximating integrals in statistics with special emphasis on Bayesian integration problems. *Statistical science*, 10(3):254–272, 1995.
16. Christian Robert and George Casella. *Monte Carlo statistical methods*. Springer Science & Business Media, 2013.
17. Constantinos Constantinopoulos and Aristidis Likas. Unsupervised learning of Gaussian mixtures based on variational component splitting. *IEEE Transactions on Neural Networks*, 18(3):745–755, 2007.
18. Hagai Attias. A variational Bayesian framework for graphical models. In *Advances in neural information processing systems*, pages 209–215, 2000.
19. Adrian Corduneanu and Christopher M Bishop. Variational Bayesian model selection for mixture distributions. In *Artificial intelligence and Statistics*, volume 2001, pages 27–34. Morgan Kaufmann Waltham, MA, 2001.
20. Mark William Woolrich and Timothy E Behrens. Variational Bayes inference of spatial mixture models for segmentation. *IEEE Transactions on Medical Imaging*, 25(10):1380–1391, 2006.
21. Michael I Jordan, Zoubin Ghahramani, Tommi S Jaakkola, and Lawrence K Saul. An introduction to variational methods for graphical models. *Machine learning*, 37(2):183–233, 1999.
22. Jeffrey Regier, Andrew Miller, Jon McAuliffe, Ryan Adams, Matt Hoffman, Dustin Lang, David Schlegel, and Mr Prabhat. Celeste: Variational inference for a generative model of astronomical images. In *International Conference on Machine Learning*, pages 2095–2103, 2015.
23. David M Blei, Alp Kucukelbir, and Jon D McAuliffe. Variational inference: A review for statisticians. *Journal of the American Statistical Association*, 112(518):859–877, 2017.
24. Christopher M Bishop. Variational learning in graphical models and neural networks. In *International Conference on Artificial Neural Networks*, pages 13–22. Springer, 1998.
25. JM Bernardo, MJ Bayarri, JO Berger, AP Dawid, D Heckerman, AFM Smith, M West, et al. The variational Bayesian em algorithm for incomplete data: with application to scoring graphical model structures. *Bayesian statistics*, 7:453–464, 2003.

26. Nizar Bouguila and Djemel Ziou. High-dimensional unsupervised selection and estimation of a finite generalized Dirichlet mixture model based on minimum message length. *IEEE transactions on pattern analysis and machine intelligence*, 29(10):1716–1731, 2007.
27. Antonio Penalver and Francisco Escolano. Entropy-based incremental variational Bayes learning of Gaussian mixtures. *IEEE transactions on neural networks and learning systems*, 23(3):534–540, 2012.
28. Wentao Fan, Faisal R Al-Osaimi, Nizar Bouguila, and Jixiang Du. Proportional data modeling via entropy-based variational Bayes learning of mixture models. *Applied Intelligence*, 47(2):473–487, 2017.
29. Wentao Fan, Nizar Bouguila, Sami Bourouis, and Yacine Laalaoui. Entropy-based variational Bayes learning framework for data clustering. *IET Image Processing*, 12(10):1762–1772, 2018.
30. W Raghupathi and S Kudyba. Healthcare informatics: improving efficiency and productivity. In *Data Mining in Health Care*, pages 211–223. 2010.
31. George G Tiao and Irwin Cuttman. The inverted Dirichlet distribution with applications. *Journal of the American Statistical Association*, 60(311):793–805, 1965.
32. D Chandler. Oxford university press; new york: 1987. *Introduction to Modern Statistical Mechanics*, pages 234–270.
33. Josef Kittler, Mohamad Hatef, Robert PW Duin, and Jiri Matas. On combining classifiers. *IEEE transactions on pattern analysis and machine intelligence*, 20(3):226–239, 1998.
34. Lev Faivishevsky and Jacob Goldberger. Ica based on a smooth estimation of the differential entropy. In *Advances in neural information processing systems*, pages 433–440, 2009.
35. Nikolai Leonenko, Luc Pronzato, Vippal Savani, et al. A class of rényi information estimators for multidimensional densities. *The Annals of Statistics*, 36(5):2153–2182, 2008.
36. WHO. *Cardiovascular Diseases report of WHO*. https://www.who.int/health-topics/cardiovascular-diseases/.
37. Chayakrit Krittanawong, HongJu Zhang, Zhen Wang, Mehmet Aydar, and Takeshi Kitai. Artificial intelligence in precision cardiovascular medicine. *Journal of the American College of Cardiology*, 69(21):2657–2664, 2017.
38. UCI repository. *Heart disease*. https://archive.ics.uci.edu/ml/datasets/Heart+Disease/.
39. WHO. *Diabetes disease fact sheet*. https://www.who.int/news-room/fact-sheets/detail/diabetes/.
40. Ioannis Kavakiotis, Olga Tsave, Athanasios Salifoglou, Nicos Maglaveras, Ioannis Vlahavas, and Ioanna Chouvarda. Machine learning and data mining methods in diabetes research. *Computational and structural biotechnology journal*, 15:104–116, 2017.
41. Kaggle. *Diabetes disease dataset*. https://www.kaggle.com/uciml/pima-indians-diabetes-database/.
42. WHO. *WHO cancer statistics*. https://www.who.int/news-room/fact-sheets/detail/cancer/.
43. UCI. *Lung cancer*. https://archive.ics.uci.edu/ml/datasets/Lung+Cancer//.
44. WHO report on breast cancer. *Breast cancer dataset*. https://www.who.int/cancer/prevention/diagnosis-screening/breast-cancer/en//.
45. breast cancer. *Cytological breast tissue dataset*. https://archive.ics.uci.edu/ml/datasets/Breast+Cancer+Wisconsin+(Diagnostic)/.

Printed in the United States
by Baker & Taylor Publisher Services